外教社 英汉·汉英百科词汇手册系列

总主编　庄智象

英汉·汉英
动力与电气工程词汇手册

An English-Chinese Chinese-English
Glossary of Power and
Electrical Engineering

编者　段宇杰　李　争　陆少兵
　　　袁健兰　干　诚

上海外语教育出版社
外教社 SHANGHAI FOREIGN LANGUAGE EDUCATION PRESS

图书在版编目(CIP)数据

英汉·汉英动力与电气工程词汇手册/段宇杰编.
—上海:上海外语教育出版社,2013
(外教社英汉·汉英百科词汇手册系列)
ISBN 978-7-5446-3373-4

Ⅰ.①英… Ⅱ.①段… Ⅲ.①动力工程-词汇-手册-英、汉 ②电气工程-词汇-手册-英、汉 Ⅳ.
①TK-62 ②TM-62

中国版本图书馆CIP数据核字(2013)第079177号

出版发行:上海外语教育出版社
(上海外国语大学内) 邮编:200083
电　　话: 021-65425300(总机)
电子邮箱: bookinfo@sflep.com.cn
网　　址: http://www.sflep.com.cn http://www.sflep.com
责任编辑: 贺　敏

印	刷:	上海信老印刷厂
开	本:	787×965 1/32 印张 8.625 字数 374千字
版	次:	2013年5月第1版 2013年5月第1次印刷
印	数:	3 000册
书	号:	ISBN 978-7-5446-3373-4 / T・0041
定	价:	20.00元

本版图书如有印装质量问题,可向本社调换

目录

序 v

前言 viii

使用说明 ix

英汉部分 1~128

汉英部分 129~260

参考文献 261

外教社英汉·汉英百科词汇手册系列

总主编:庄智象

英汉·汉英动力与电气工程词汇手册

编　　者:段宇杰　李　争　陆少兵　袁健兰　干　诚
策划编辑:张春明　李法敏
责任编辑:贺　敏

序

改革开放 30 年来,我国的经济、科技、文化、教育、外贸和社会各项事业快速发展,取得了令人瞩目的成就,经济总量翻了两番之多,国内生产总值 2007 年已近 25 万亿元人民币,人均国民收入达到了 2 000 多美元,提前实现了小康目标;科技水平不断提高,高新技术快速发展,缩短了与世界先进国家的差距;文化、教育事业迅猛发展,不断满足和丰富着人民日益增长的文化精神需要;社会服务和保障体系不断完善,使中华民族和社会更加和谐。我国在国际事务中的作用日益凸显,与世界各国的政治、经济、科技、文化、教育、外交、军事等方面的交往日益频繁。成功加入世界贸易组织,成功申办和举办 2008 年北京奥运会和残奥会,成功申办上海世博会等一系列重大外交活动,塑造了中国在世界上的良好形象,更是促进了我国与世界各国的交往、交流和交融。经济全球化、科技一体化、文化多元化、信息网络化的发展趋势,使人们的生活正发生着深刻的变化。

新的学科、新的术语、新的词汇不断诞生和涌现。我国数亿不同层次的英语学习者和使用者,在学习或工作中不断遇到新的词语、新的术语。大部分的英语学习者和使用者能够比较顺利地用普通英语进行交流和交际,而一旦遇到专业领域的词汇或术语往往会陷入困境,有时可能了解某词语的一般意义,但不知道它在某些专业领域指的是什么。这种情况,在日常交往中,或在专业性较强的国际会议中屡见不鲜,常常需要英语专家和专业方面的人士通力合作,才能解疑释惑。若能编纂出版一套英汉、汉英的百科词汇手册,则将大大有益于英语学习者和使用者,能够为他们的英语学习和使用助上一臂之力。笔者曾在上世纪90年代初随同中国教材出版项目考察团访美,任翻译,就是依靠了一本英汉汉英出版印刷词汇手册,比较顺利地完成了任务。同时,此类专业词汇手册还有助于学习者提高英语水平和能力,借助英语汲取科技知识和信息,扩大视野,不断丰富和提高专业知识和水平。有鉴于此,上海外语教育出版社(以下简称外教社)与牛津大学出版社合作,于本世纪初陆续出版了牛津百科分类词典英语版40余种。这批百科词典的出版深受专业人员、英语学习者和使用者的欢迎。同时,根据部分读者的要求,外教社经过充分调研和论证,并征得牛津大学出版社同意,从该系列词典中挑选出近十种,组织有关专业人员编译成英汉双解版;由于工作量巨大,亦仅将这部分词典的词目翻译成汉语。随着对外交流和交往的深入发展,不断有读者询问外教社是否出版有完整的百科方面的词汇手册或词典。为满足这一需要,外教社经过广泛的调研、需求分析和市场论证,组织编写了外教社英汉·汉英百科词汇手册系列图书,历经四年的努力,全国近百位编纂者的通力合作和辛勤劳动,终于迎来了第一批书稿。

本百科词汇手册系列有以下几个特点:

1. 按学科、专业和行业分册编写(以下统称专业),分类依据主要是国家标准学科分类、国家标准国民经济行业分类、企业经营行业分类及中国图书在版编目分

类,涵盖自然科学、技术、社会科学、人文科学、社会生活等 80 余个领域;

2. 各专业词汇手册包括英汉、汉英两部分,尽可能涵盖各专业最基本、最常用的词汇,每部分收词基本上控制在 5 000 至 10 000 条,版式和开本设计便于使用和携带;

3. 各分册按词汇的使用频率收列专业基本词汇,同时力求反映该专业的最新发展,只收专业性词汇和普通语文词汇的专业性义项。与本专业关系不大的其他专业词汇尽量不收。

本系列词汇手册可供英语学习者、使用者及相关专业人员了解和熟悉专业词汇,学习和丰富专业知识,提升专业视野和水平之参考,亦可作为翻译的参考工具书。由于学识和能力所限,新的词汇层出不穷,收词难免挂一漏万,谬误和缺陷在所难免,敬请广大读者惠予批评、指正。

庄智象
2008 年 10 月

前 言

早在16世纪,人类就开始研究电磁现象。后来随着发电机、电动机的发明、输配电技术的不断发展,电能逐渐成为现代社会不可或缺的能源。电气工程涉及工农业生产和人类生活的方方面面,在国民经济发展中发挥着至关重要的作用。动力工程是电气工程的分支学科,可持续地开发、利用能源与人类社会的未来息息相关。近些年,动力与电气工程学科快速发展,新的术语和专业词汇不断增加,在这种背景下,我们本着专业、实用的指导思想,编写了这本《英汉·汉英动力与电气工程词汇手册》。

本手册汇集了电气工程、动力工程及相关领域的常用词汇,以及反映这些领域技术发展和科学成果的新术语。本书双向收词16 000余条,涉及工程热物理、热工学、动力机械工程等领域,包括工程传热、传质学、燃烧学、热工测量与仪器仪表、制冷工程、供热工程、内燃机工程、电工学、电机学、电气学、电力电子技术、高电压工程、绝缘技术、发电工程(水力、风力、热力、磁流体发电)、输配电工程、电力系统与自动化等学科。本书旨在帮助大专院校或科研院所中的动力与电气工程专业师生、研究者及技术人员学习专业英语、研读专业英语资料,同时也可供相关人员从事专业翻译参考之用。

本手册由上海电机学院的段宇杰、李争、陆少兵、袁健兰、干诚等共同编写。在编写过程中,我们得到了复旦大学、上海交通大学、同济大学、上海电力学院、上海电机学院、上海应用技术学院、上海电气集团、上海外语教育出版社等高校、企业单位和出版界的专家、学者的支持和帮助,在此谨一并表示衷心的感谢!

由于编者水平有限,书中疏漏之处在所难免,敬请广大读者批评指正,提出宝贵意见。

使 用 说 明

1. 本手册分为英汉、汉英两部分。
2. 英汉部分条目按照字母顺序排列;汉英部分条目按照拼音音序排列,同音字按照笔画数排列。
3. 英汉部分以数字、特殊符号、希腊字母等非英语字母开头的条目放在英汉部分正文之后;汉英部分以数字、特殊符号、西文字母等非汉字开头的条目放在汉英部分正文之后。
4. 外国人名条目的词头与译文一般采用姓前名后的形式。例如:

    Chomsky, Avram Noam 乔姆斯基,A·N

    乔姆斯基,A·N Chomsky, Avram Noam
5. 为便于查阅,汉英部分的某些条目(多为自由组合)以汉语中心词或关键词立目,其他文字置于空心鱼尾号中,放在中心词或关键词之后。例如:

    与产品相关的行为变量　product-related behavioural descriptor

    立目时处理为:

    行为变量〖与产品相关的〗 product-related behavioural descriptor
6. 英汉部分概念相同或相近的汉语译文之间用逗号分隔,概念不同的汉语译文之间用分号分隔;汉英部分的英语译文一律用分号间隔。
7. 圆括号"( )"用于括注:① 缩写或全称形式;② 可省略的内容;③ 解释说明性文字。
8. 方括号"[ ]"用于括注可替换的文字。
9. 尖括号"〈 〉"用于括注语言及学科标注。

英汉部分

AAC (automatic amplitude control) 自动幅值控制
abbertite 黑沥青
ablative 烧蚀的
abrasion 磨损,磨耗
abrasivity 冲蚀度,磨蚀度
ABS (acrylonitrile-butadiene-styrene) 丙烯腈-丁二烯-苯乙烯(共聚物)
abscissa 横坐标
absolute error 绝对误差
absolute humidity 绝对湿度
absolute pressure 绝对压强,绝对压力
absolute temperature 绝对温度
absolute zero 绝对零度
absorbent 吸附剂,吸收剂
absorber 吸收材料;减震器,缓冲器
absorption refrigeration 吸收制冷
abstraction 提取,抽取;抽象(化)
abutment 拱座;桥台
AC (alternating current) 交流电流,交流
ACC (automatic combustion control) 自动燃烧控制,燃烧自动调节
accessory 附件

accumulator 存储器;累加器;蓄电池;蓄压器;安全注射箱
ACE (automatic computing equipment) 自动计算装置
acidimetry 酸量(滴定)法
ACP (auxiliary control panel) 辅助控制盘
acrylonitrile-butadiene-styrene (ABS) 丙烯腈-丁二烯-苯乙烯(共聚物)
ACSR cable (aluminium conductor steel-reinforced cable) 钢芯铝线
activator 活化剂,活性剂,激活剂
active circuit 有源电路
active current 有功电流
activity (放射性)活度,活性
actuator 操动件;驱动器;执行器
adamant 硬石
adaptor 结合器,适配器,连接器
ADC (analogue-digital converter) 模数转换器
addendum 齿顶高
additive 添加剂,添加物
A-derrick 人字起重机
adhesion 黏附,黏合,附着;附着力
adhesive 黏合剂,胶黏剂
adjuster rod 调节棒

adjustment 调节,调整;找正
adjutage 喷射管,放水管
admixture 掺和料,掺和物;外加剂
adobe 土坯砖;砖坯黏土
adsorption 吸附(作用)
adze 扁斧,锛子
AE (air entraining) 加气,掺气
aeolipile 汽转球
aerated concrete 加气混凝土
aeration 曝气,充气
aerator 曝气器,充气器
aerosol 气溶胶;悬浮微粒;气雾剂
AFC (automatic following control) 自动跟踪控制
afterburner 加力燃烧室,补燃室
afterburning 加力燃烧,补燃
after-condenser 后冷凝器
aftercooler 后冷器
AFWC (automatic feed water control) 自动给水控制
AGC (automatic generation control) 自动发电控制
agglomerant 凝聚剂
agglomerate 聚集;团粒,团块
agitation 搅拌,搅动
agitator 搅拌器
AH (air preheater) 空气预热器
Ah (ampere-hour) 安时,安培小时
aiguille 钻头,钻孔器
air-breather 通气孔,通气装置
air-cooler 空气冷却器
air-cooling 空气冷却
air entraining (AE) 加气,掺气
air gap 气隙,空气(间)隙
air gauge 气压计
air intake 进风;进气道
airlock 锁气器;气闸
air preheater (AH) 空气预热器
airway 通气管;航路
AIS (alarm indication signal) 报警指示信号
Alclad 包层铝
alkali 碱

alkalinity 碱度
alkalinization 碱化(作用)
allocation 分配
allocator 分配器
alloy 合金
alluvium 冲积层
all volatile treatment (AVT) 全挥发性处理,零固形物处理
alphabetic code 字母码
alpha radiation α射线,阿尔法射线
alternating current (AC) 交流电流,交流
alternator 交流发电机
alum 明矾;硫酸铝
aluminium 铝
aluminium bronze 铜铝合金,铝青铜
aluminium conductor steel-reinforced cable (ACSR cable) 钢芯铝线
aluminium oxide 三氧化二铝,氧化铝
ammonia 氨
ammonification 氨化作用
ammoniometer 氨量计
ammonium 铵
ammonium chloride 氯化铵
ampacity 载流量
ampere-hour (Ah) 安时,安培小时
ampere-meter 安培表,电流表
amplification 放大
amplifier 放大器
amplitude 振幅;幅度
anabolism 合成代谢
analogue 类似物
analogue circuit 模拟电路
analogue-digital converter (ADC) 模数转换器
analogue-to-digital converter 模数转换器
ancillary equipment 辅助设备
ancon 悬臂托梁
anemometer 风速计
angle 角;角度
anion 阴离子

anisotropy 各向异性
annealing 退火
anode 阳离子
anthracite 无烟煤,硬煤
antiknock 抗爆剂
apparatus 机器;仪器,设备
application reference data (ARD) 应用参考数据
application specific integrated circuit (ASIC) 专用集成电路,特定用途集成电路
APT (Automatically Programmed Tool) APT 语言
aqua ammonia 氨水
aqua thruster 气压扬水机
aqueous solution 水溶液
ARC (automatic remote control) 自动遥控
arc 弧;电弧
arc-back 逆弧
arc energy 电弧能
arc initiation 引弧
arc lamp drive 弧光灯驱动器
arc melting 电弧熔化
arc welding 电弧焊,弧焊
ARD (application reference data) 应用参考数据
ardometer 辐射高温计,光测高温计
argon 氩
arm 臂
armature 电枢,转子;衔铁
arm-brace 撑脚
arrester 制动器;捕捉器,搜集器
articulation 铰接;清晰度
aseptic 无菌的;防腐剂
ash pit 灰坑
ASIC (application specific integrated circuit) 专用集成电路,特定用途集成电路
assembler 装配器;汇编程序
assembly 装配;组件,组合装置
asymmetry 不对称度,不平衡度
ATC (automatic turbine control) 汽轮机自动控制
atom 原子

atomic battery 原子电池
atomic collision 原子碰撞
atomic composition 原子组成
atomic fission 原子裂变
atomic number 原子序数
atom smasher 粒子加速器
attemperator 减温器
attenuator 衰减器
attrition 磨损
audio-circuit 声频电路
auger 螺旋钻
automated information processing 自动信息处理
automated mapping 自动制图
Automatically Programmed Tool (APT) APT 语言
automatic amplitude control (AAC) 自动幅值控制
automatic breaker 自动断路器
automatic combustion control (ACC) 自动燃烧控制,燃烧自动调节
automatic computing equipment (ACE) 自动计算装置
automatic feed water control (AFWC) 自动给水控制
automatic following control (AFC) 自动跟踪控制
automatic generation control (AGC) 自动发电控制
automatic measuring system 自动测量系统
automatic remote control (ARC) 自动遥控
automatic sequence (操作过程的)自动顺序
automatic turbine control (ATC) 汽轮机自动控制
autotransformer 自耦变压器
autotransformer starter 自耦变压器起动器
auxiliary circuit 辅助电路,辅助回路
auxiliary control panel (ACP) 辅助控制盘
auxiliary report 辅助报告

auxiliary winding 附加绕组
AVT (all volatile treatment) 全挥发性处理,零固形物处理
axial fan 轴流式通风机

axial focusing 轴向聚焦
axis-flow 轴流式
axonometry 轴测法

backbone 主干
backplan 底视图
back siphonage 反虹吸
badigeon 油灰
baffle 挡板,围板
balance of plant (BOP) 电厂配套设施
ballast 镇流器
ballast factor 镇流器系数
ball valve (BV) 球阀
band 波段;频带
bar¹ 半线圈,线棒;杆
bar² 巴(压强单位)
barium 钡
baroceptor 气压传感器
barometer 气压计
barometric pressure 大气压力
barothermograph 气压温度计
barycentre 重心,质心
base 灯头;管座;基极;碱
base coordinate system 基座坐标系
baseline 基线
basic insulation level (BIL) 基本绝缘水平
basicity 碱度
batcher 计量送料器
batten 板条,挂瓦条
baud 波特(数据发送速率的单位)
bauxite 铝土矿
bayonet base 卡口灯头
bayonet cap 卡口灯头
BCD (binary-coded decimal) 二进码十进数
BCT (bushing current transformer) 套管式电流互感器

beacon 灯标;信标
bead 焊珠;墙角圆
beam-compasses 长臂圆规
beam splitter 分束器
beard 倒钩
bearer 载体;支座,托架
bearing 方位;轴承
becket 索环,绳环
bed 层;砂床
bedplate 底板
bellying 膨胀
belt conveyor 皮带运输机
benchmark 基准
bending strength 弯曲强度
bent 排架,横向构架
benzene 苯
BER (bit error ratio) 误比特率,位误差率
bevel 斜角,斜面;倾斜
BFE (boiler-front equipment) 炉前(点火控制)设备
BFP (boiler feed pump) 锅炉给水泵
BF system (boiler feed system) 锅炉给水系统
BFV (butterfly valve) 蝶阀,蝶形阀
BHN (Brinell hardness number) 布氏硬度数
bias 偏压;偏置;偏离
biaxial stretching 双向拉伸
bicarbonate 碳酸氢盐,重碳酸盐
biconcave 双面凹的
biconvex 双面凸的
bifurcation 分叉
BIL (basic insulation level) 基本绝缘水平

bimetal 双金属
bimetallic corrosion 双金属腐蚀
bimoment 双力矩
binary 二进制
binary-coded decimal (BCD) 二进码十进数
biocycle 生物环
biodegradation 生物降解
bi-phase 双相的
bit¹ 钻头
bit² 比特,二进制位
bit error ratio (BER) 误比特率,位误差率
bitumen 沥青
bituminite 沥青质体
bituminous coal 烟煤
BL (boundary line) 边界线
blast 鼓风
bleach 漂白剂
bleeder 放散管
blemish 缺陷,瑕疵
blister 气泡;砂眼
block diagram 框图,方块图
blowdown 排出
blower 送风器,鼓风机
blow-off 放气
bogie 转向架
boiler feed pump (BFP) 锅炉给水泵
boiler feed system (BF system) 锅炉给水系统
boiler-front equipment (BFE) 炉前(点火控制)设备
boiler vessel 锅炉体
bolection 凸嵌线
bolometer 辐射热(测量)计
bolster 垫块,支撑物;托木
bolt 螺栓;插销
bolter 筛粉机
bombardment 轰击
bonnet 护罩,护盖
boom 悬臂
booster 升压机;助推器
booster transformer 增压变压器,吸流变压器

BOP (balance of plant) 电厂配套设施
bore 口径
boron 硼
bounce 反弹
boundary line (BL) 边界线
brake 制动器,刹车,闸
brass 黄铜
breadboard 实验电路板
bridge crane 桥式起重机,桥吊
bridle 束带
brightness 亮度
brine 盐水,卤水
Brinell hardness number (BHN) 布氏硬度数
British Standards Institution (BSI) 英国标准协会
British thermal unit (Btu) 英(制)热单位
brittleness 脆性
bromide 溴化物
brown-out 局部停电
brush 电刷
BSI (British Standards Institution) 英国标准协会
Btu (British thermal unit) 英(制)热单位
buffer 缓冲器,减震器
bulk supply 批量供应
buoyancy 浮力
burden 负荷
burette 滴定管,量管
burner 燃烧器,喷燃器
burn-in 老化;预烧测试
burn-up 燃耗
busbar 母线,汇流排
bushing current transformer (BCT) 套管式电流互感器
butane 丁烷
butt contact 对接接触;对接触头
butterfly valve (BFV) 蝶阀,蝶形阀
butt weld (BW) 对接焊缝
BV (ball valve) 球阀
BW (butt weld) 对接焊缝

cab (cellulose acetate butyrate) 醋酸丁酸纤维素
cabinet 机箱
CAD (computer-aided design) 计算机辅助设计
cadmium 镉
caking 结块
calcium 钙
calibration 校准,标定;刻度
calibrator 校准器
calibre 管径,口径
caliper 卡尺;卡钳
calorie 卡
calorimeter 量热器,热量计
camber 起拱,弯曲度,侧向弯度
cantilever 悬臂梁,伸臂
CAP (computer-aided production) 计算机辅助生产
capacitance 电容
capacitor 电容器
capacitor voltage transformer (CVT) 电容式电压互感器
capacity 容量;能容
carbide 碳化物
carbon 碳
carbonate 碳酸盐
carbonite 天然焦
carbonization 碳化,干馏
carbon rod 碳棒
carbon steel 碳(素)钢
carboxylic acid 羧酸
carburettor 汽化器
carcass 构架,框架
cartridge 熔管;套筒
cast iron 铸铁
catalysis 催化
catalyst 催化剂

catcher 收集器,捕捉器
catcher resonator 获能腔
catenary 悬链线
cathode 阴极,负极
cathode ray tube (CRT) 阴极射线管
cation 阳离子,正离子
cation exchange resin 阳离子交换树脂
cavitation 空化;气穴,气蚀
cavity 空泡;空隙
CCCW (closed cycle cooling water) 闭式循环冷却水
CLCS (closed loop control system) 闭环控制系统
CCW (counterclockwise) 逆时针
cellular structure 胞状组织,微孔结构
celluloid 赛璐珞
cellulose acetate butyrate (cab) 醋酸丁酸纤维素
Celotex board 隔音板
CEMS (continuous emission monitoring system) 连续排放监测系统
central processing unit (CPU) 中央处理器
centrifugal compressor 离心式压气机
centrifugation 离心分离
centrifuge 离心机
ceramics 陶器
cetane 十六烷
CFB (circulating fluidized bed) 循环流化床
CFC (chlorofluorocarbon) 氟氯化碳

chang-over switch (COS) 转换开关
charcoal 炭
charge 充电;炉料;荷
chemiluminescence 化学发光
chemosynthesis 化能合成
chloramine 氯胺
chloride 氯化物
chloride ion 氯离子
chlorinated polyether (CPE) 氯化聚醚
chlorinated polyvinyl chloride (CPVC) 氯化聚氯乙烯
chlorination 氯化;加氯处理
chlorinator 氯化器
chlorine 氯
chlorofluorocarbon (CFC) 氟氯化碳
chopper 斩波器
circuit 电路,回路
circuit-breaker 断路器
circuit-breaker capacitor 断路器电容器
circulating fluidized bed (CFB) 循环流化床
circulation 循环;环流;环流量
circulator 循环器;环行器
clad temperature 包壳温度
clamping device 夹件
clapotis 驻波,立波
classifier 分类器;粗粉分离器
clastic 碎屑
clevis 帽槽;U形钩
clinometer 测角器,倾斜仪
closed cycle cooling water (CCCW) 闭式循环冷却水
closed loop control system (CLCS) 闭环控制系统
clutch 离合器
coagulation 凝固,凝聚;聚沉
coal tar 煤焦油,煤沥青
coaxial cable 同轴电缆
cock 旋塞阀
code (代)码
code converter 代码转换器
coefficient 系数

coffin 屏蔽容器
cogeneration 热电联产(尤指利用工业余热发电)
cohesion 内聚力;内聚性
coil 绕组,线圈;盘管
collapsibility 湿陷性;溃散性
collar 轴环,接头
collector 集电极;捕收剂
collimator 准直仪,准直管
colorimeter 比色计,色度计
colorimetry 比色法,色度测量;色度学
colority 色度
combimeter 多功能电能仪表
combined voltage regulation 混合调压
combiner 合路器
combustibility 可燃性
combustion 燃烧
combustor 燃烧器;燃烧室
commingler 混合器
comminutor 粉碎机
commissioning test 投运试验
commissure 接缝,接合处;焊接处
commutator 换向器;交换子
compaction 压实,夯实
compatibility 兼容性,相容性
complementarity 互补性
complete quadratic combination (CQC) 完全二次型方根组合法
composite error 复合误差
compression ignition 压缩点火
compression stroke 压缩冲程
compressor 压缩机,压气机
computer-aided design (CAD) 计算机辅助设计
computer-aided production (CAP) 计算机辅助生产
concentrate 浓缩物
concentrator 集中器;集线器
concentric coil 同心式线圈
condensate 凝结水;冷凝物
condensate pump (CP) 凝(结)水泵

condenser 凝汽器,冷凝器;电容器;聚光器
conditionality 制约性
conditioner 调节器;调节剂
conductance 电导
conductibility 导电性
conductivity 导电性;电导率
conductometer 电导仪,电导计
conduit 电线管;输水管道
configuration 配置;组态;构型
congelation 冻凝,冻结
congestion 阻塞,拥塞
connection 连接;联结
connector 连接器;接头
console 控制台,操纵台
constant 常数,常项,常量
constant flux voltage regulation 恒磁通调压
constant pressure 定压,等压
constant pressure start-up 定压启动
contactless relay 无触点继电器
containment 安全壳;封闭
contaminant 污染物,杂质
contamination 污染
content 含量;库容
contingency 偶然性;意外事故,意外事件
continuity 连续性,持续性
continuous emission monitoring system (CEMS) 连续排放监测系统
contour 等高线;周线
contour interval 等高距
control accuracy 控制准确度,控制精(确)度
control apparatus 控制电器
control box 控制箱
control circuit 控制电路
control contact 控制触头
control electrode 控制极
control exciter 控制励磁机
control panel 控制盘
control switch (CS) 控制开关
control transformer 控制变压器
control wiring diagram (CWD) 控制接线图
convection 对流
convector 对流器
converter 转换器;变流器,换流器
converter transformer 变流变压器,换流变压器
convex 凸
convolution 卷积
coolant 冷却剂,冷却介质
coordinate 坐标
copolymer 共聚物
copper 紫铜,铜
copper sulphate 硫酸铜
cord 电线,软线;索
core 杆体;线芯,堆芯;铁芯
cork 软木塞
corona 电晕
corrector 校正器;校正子
correlation diagram 相关图
corrosion-resistant steel 不锈钢
corrosive 腐蚀性物质;腐蚀的
corrugated steel sheet 波纹(钢)板,瓦垅板
COS (change-over switch) 转换开关
cosecant 余割
cosine 余弦
counterclockwise (CCW) 逆时针
coupler 连杆;耦合器
coupling 联管节;联轴器,联轴节;耦合
cowl 通风帽;整流罩
CP (condensate pump) 凝(结)水泵
CPE (chlorinated polyether) 氯化聚醚
CPU (central processing unit) 中央处理器
CPVC (chlorinated polyvinyl chloride) 氯化聚氯乙烯
CQC (complete quadratic combination) 完全二次型方根组合法
crab 起重机,吊车
crankshaft 曲轴

crawler 履带式车辆
CRC (cyclic redundancy check) 循环冗余校验
creepage discharge 爬行放电
creepage distance 爬电距离
crest factor 峰值因数
crib 枕盒
critical point 临界点
cross bar 纵横开关,交叉棒
cross-connection 交叉连接
cross curvature 横向曲率
cross head 十字头
cross-section 横断面,横截面
crosstalk 串音;串扰
CRT (cathode ray tube) 阴极射线管
cryogenics 低温学
cryopedology 冻土学
crystal 结晶(体);晶体
CS (control switch) 控制开关
CT¹ (current transformer) 电流互感器
CT² (current transmitter) 电流变送器
cumulant 累积量
current 电流;流
current element 电流元
current error 电流误差
current injection 电流引入
current limiter 限流器

current meter 流速仪
current recovery ratio 电流恢复比
current relay 电流继电器
current transformation ratio 变流比
current transformer (CT) 电流互感器
current transmitter (CT) 电流变送器
cursor 光标
curvature 弯曲;曲率
curve 曲线
cushion 衬垫;缓冲垫
CVT (capacitor voltage transformer) 电容式电压互感器
CWD (control wiring diagram) 控制接线图
cybernetics 控制论
cyclane 环烷烃
cyclic redundancy check (CRC) 循环冗余校验
cycloconverter 周波变流器,循环变频器
cyclone 气旋;旋风分离器
cylinder 筒;汽缸;钢瓶
cylindrical valve 圆筒形阀
cymometer 频率计,波长计

DAC (digital-to-analogue converter) 数模转换器
D/A conversion (digital-to-analogue conversion) 数模转换
D-action 微分作用，D作用
daf basis (dry ash-free basis) 干燥无灰基
damper 风门；阻尼器；减震器
damping 阻尼；衰减
data circuit terminating equipment (DCTE) 数据电路终接[端接]设备
data terminal equipment (DTE) 数据终端设备
DC (direct current) 直流电
D-controller 微分控制器，D控制器
DCTE (data circuit terminating equipment) 数据电路终接[端接]设备
DDF (digital distribution frame) 数字配线架
deaerator 除氧器
decay 衰变；衰减
decibel 分贝
decimal system 十进制
decipherer 译码器
decoder 解码器
decomposition 分解
decompressor 减压器
decontamination 去污
de-energize 断电
de-excitation 退激
deflagration 爆燃
deflector 折向器；偏转板
deflocculant 解絮凝剂
deflocculation 反絮凝；解絮凝

degasifier 除气器
degradation 降级；降解
dehumidifier 除湿器，去湿器
dehydrant 脱水剂
dehydration 脱水(作用)
deionization 消电离
deload 减负荷
demethanizer 脱甲烷塔，甲烷馏除器
demineralization 脱盐，除盐
demultiplexer 分用器，多路分配器
densimeter 比重计，密度计
density 密度
deoxygenation 脱氧
depentanizer 戊烷馏除器，脱戊烷塔
depletion layer 耗尽层
derivative action 微分作用，D作用
derivative controller 微分控制器，D控制器
derivometer 测偏仪
derrick 抱杆；动臂起重机；井架
desiccant 干燥剂
design condition 设计工况
design head 设计水头
desulphurization 脱硫
desulphurization during combustion 燃烧脱硫
desuperheater 减温器
detector 探测器
detonation 爆燃
devaporizer 蒸发冷却器
dewing 结露
DI (digital input) 数字输入
diameter 直径

diaphragm 隔板；膜片
die casting 压铸，压力铸造
dielectric 电介质，介电体
dielectric constant 介电常数
differential 差；微分
differential pressure 差压
differentiator 微分器
diffraction 衍射
diffuser 扩风器；扩压器；扩散器
diffusion 扩散；漫射
diffusivity 扩散率
digital distribution frame (DDF) 数字配线架
digital electro-hydraulic system 数字电液控制系统
digital input (DI) 数字输入
digital pulse duration modulation (DPDM) 数字脉冲宽度调制
digital-to-analogue conversion (D/A conversion) 数模转换
digital-to-analogue converter (DAC) 数模转换器
dilatability 膨胀性
dilatancy 胀塑性
dilatation 膨胀
diluent 稀释剂
diode 二极管
diode-transistor logic (DTL) 二极管-晶体管逻辑
diode valve 二极管阀
dioxide 二氧化物
diphaser 二相发电机
dipole 偶极子双极子
direct current (DC) 直流电
direct current motor 直流电动机
directional relay 方向继电器
directrix 准线
discharge 排放；放电
discharge head 排出压头
discharger 排放装置；放电器
discharging current 放电电流
disconnector 隔离开关，隔断刀闸
disinfectant 消毒剂
displacement 位移；置换；排量
dissipation 耗散
dissolution 溶解

dissolved oxygen 溶解氧
distributed mixing burner (DMB) 分布式混合燃烧器
distribution automation 配电自动化
distribution management system (DMS) 配电管理系统
distribution transformer 配电变压器
distributor box 分配箱
divergence 散度；发散
DMB (distributed mixing burner) 分布式混合燃烧器
DMS (distribution management system) 配电管理系统
dose 剂量，用量
dose equivalent 剂量当量
dosimeter 剂量计
downcomer 下降管
downwind WTG 下风式风电机
DPDM (digital pulse duration modulation) 数字脉冲宽度调制
drainage 排水
drainage pipe 排水管
drain valve 疏水阀
drawing 图
drift 漂移
drip-proof type 防滴式
drive 传动
driven pulley 从动带轮
driver 激励器；驱动器
droplet entrainment 液滴夹带
drum 汽包，锅筒；鼓轮
drum stand 绕线架，线轴
dry ash-free basis (daf basis) 干燥无灰基
dryness 干度
DTE (data terminal equipment) 数据终端设备
DTL (diode-transistor logic) 二极管-晶体管逻辑
ductility 延性，延度
ductwork 管道系统
duplex 双工
duplex lap winding 双叠绕组

duplex RTD (duplex resistance temperature detector) 双支热电阻
duplex wave winding 双波绕组
durability 耐久性,耐用性
duration 持续时间
duty 功率;工作制

dynamic load 动态负荷
dynamics 动力学
dynamic stability 动态稳定
dynamo 发电机;电动机
dynamometer 功率计
dyne 达因(力的单位)

EAL (electromagnetic amplifying lens) 电磁放大透镜
early failure 早期事故
earth 地;接地
earth bar 接地棒
earth bus 接地母线
earth capacitance 对地电容
earth circuit 接地电路
earth clamp 接地线夹
earth clip 接地夹
earth conductivity 大地电导率
earth conductor 接地导线
earth connection 接地线;接地
earth current (EC) 泄地电流,大地电流
earthed voltage 接地电压
earth electrode 接地(电)极,接地体
earth fault 接地故障
earth-fault current 接地故障电流
earth-fault protection 接地故障保护
earth-fault relay 接地(故障)继电器
earth grid 接地栅极
earth induction 地磁感应
earth inductor 地磁感应器
earthing 接地
earthing brush 接地电刷
earthing conductor 接地导体
earthing contact 接地触点
earthing reactor 接地电抗器
earthing relay 接地继电器
earthing switch 接地开关
earthing system 接地系统
earthing transformer 接地变压器
earth insulation 对地绝缘
earth jack 接地插孔
earth lead 接地导线
earth leakage 接地漏电,对地泄漏
earth leakage circuit-breaker (ELCB) 接地漏电断路器
earth leakage current 漏地电流
earth leakage fault 对地漏电故障
earth leakage relay 漏地电流继电器
earth loop 接地回路
earth magnetic field 地磁场
earth magnetism 地磁
earthometer 接地测量仪;兆欧表;高阻表
earth plate (EP) 接地板
earth plate resistance 接地板电阻
earth point 接地点
earth potential 地电位,地电势
earth potential difference (对)地电位差
earth potential working 地电位作业
earth resistance 接地电阻,大地电阻
earth resistivity 大地电阻率
earth resistor 接地电阻
earth return (ER) 大地回路,接地回线
earth return circuit 接地回流电路,地回电路
earth rod 接地棒
earth shield 接地屏蔽
earth strip 接地带
earth switch 接地开关

earth wire 地线
EAX (electronic automatic exchange) 电子自动交换机
e-beam 电子束
ebonite 硬橡胶
EBR (electron beam recorder) 电子束记录仪
EBS system (electron beam scanning system) 电子束扫描系统
EC (earth current) 泄地电流,大地电流
ECAP (electronic circuit analysis program) 电子电路分析程序
ECC (electron coupling control) 电子耦合控制
eccentric compression 偏心受压,偏心压缩
eccentricity 偏心距;离心率
eccentric load 偏心载荷
eccentric mechanism 偏心轮机构
eccentric position 偏心位置
ECDM (electrochemical discharge machining) 电化学放电加工
echelle grating 中阶梯光栅
echo 回送;回声;回波
echo attenuation 回波衰减
echo box 回波箱
echo eliminator 回波消除器
echogram 回声测深图;回声深度记录
echograph 音响测深自动记录仪
echo-image 回波像,重影
echolocation 回声定位
echometer 回声仪
echo sounder 回声测深仪
echo trap 回波抑制器
ECM (electronic countermeasures) 电子干扰,电子对抗
eddy 涡,旋涡
eddy current 涡流
eddy current braking 涡流制动
eddy current circuit 涡流电路
eddy current damping 涡流阻尼
eddy current loss 涡流损耗
eddy current motor 涡流电动机

eddy diffusion 涡流扩散
eddy velocity 涡动速度
edge effect 边缘效应
edge emission 边(缘)发射
edge frequency 边界频率
edge irregularity 边缘不规则性
Edison base 螺旋灯座
Edison battery 镍铁电池
Edison effect 爱迪生效应,热电放射效应
Edison socket 螺口灯座,螺旋式灯头
EDM (electro-discharge machining) 放电加工,电火花加工
EDP (electric diffusing process) 放电电渗处理
effect 效应
effective cross-section 有效截面
effective current 有效电流
effective impedance 有效阻抗
effective inductance 有效电感
effective load 有效负载
effective output 有效输出(功率)
effective value 有效值
effective voltage 有效电压
effect of capacitance 电容效应
effect of inductivity 感应率效应
efficiency 效率
efficiency diode 高效率二极管
effluent 废水
EG (electronic guidance) 电子制导
EHC (electrohydraulic control) 电液控制
EHF (extremely high frequency) 极高频
EHT (extra-high tension) 超高压,极高压
EHV (extra-high voltage) 超高压,极高压
EI (external insulation) 外(部)绝缘
eigenfrequency 本征频率
eigenvalue 本征值
ejector 喷射泵;引射器

EL (electroluminescence) 电致发光
elastance 倒电容
elastic axis 弹性轴
elastic collision 弹性碰撞
elasticizer 增弹剂
elastic modulus 弹性模量,弹性模数
elasticoviscosity 弹黏性
elastic store 缓冲存储器
elastivity 倒介电常数,介电常数倒数
elbaite 锂电气石
elbow 弯管,弯头,弯管接头
ELCB (earth leakage circuit-breaker) 接地漏电断路器
electrical damper 电气阻尼器
electrical degree 电度
electrical dynamometer 电动测功机
electrical resistance welding (ERW) 电阻焊
electrical zero (EZ) 电零,电零点
electrical zero adjuster 电零位调节器
electric charge 电荷
electric conduction 电气传导
electric conductivity 电导率
electric contact 电接触
electric diffusing process (EDP) 放电电渗处理
electric dipole 电偶极子
electric field 电场
electric field intensity 电场强度
electric field line 电场线,电力线
electric flux 电通(量)
electric flux density 电通密度
electric hysteresis 电滞
electric induction 电感应
electricity 电气
electricity meter 电量计
electric potential difference 电位差
electric power storage (EPS) 蓄电池
electric power supply (EPS) 电力电源
electric propulsion (EP) 电(力)推进
electric resonance 电共振
electric shielding 电屏蔽
electric shock 电击,触电
electric soldering 电焊
electric spark igniter 电火花点火器
electric tension 电压
electrino 电中微子
electrobalance 电动天平
electrochemical discharge machining (ECDM) 电化学放电加工
electrocorrosion 电腐蚀
electrode 电极
electrode arm 电极握臂,焊条夹
electrode bias 电极偏压
electrode capacitance 电极电容,极间电容
electrode composition 电分解作用
electrode conductance 电极电导
electrode edge 电极端
electrode gap 电极间隙
electrodeless discharge 无电极放电
electrodeposition 电沉积
electrode potential 电极电位,电极电势
electrode wall 电极壁
electrodialysis 电渗析,电透析
electro-discharge machining (EDM) 放电加工,电火花加工
electrodispersion 电分散作用
electrodissolution 电解溶解
electrodrill 电动钻具
electroduster 静电喷粉器
electrodusting 静电喷粉
electrodynamic force 电动力
electrodynamic potential 电动势
electrodynamic relay 电动式继电器
electrodynamics 电动力学
electrodynamometer 力测电流计,电力测功计,电测力计

electrodynamometry 电力测功法
electroendosmosis 内电渗
electro-equivalent 电化当量
electroetching 电蚀刻
electroextraction 电解萃取,电析
electrofacing 电镀
electro-forging 电锻
electroforming 电铸
electrogen 光电分子
electrogilding 电镀金
electrogoniometer 电测角器,电测角仪
electrograph 电位记录器
electrograving 电蚀刻
electrogravity 电控重力
electrogravure 电刻
electrohydraulic actuator 电液执行机构
electrohydraulic control (EHC) 电液控制
electrohydraulic forming 水中放电成形,电液成形
electro-hydrometallurgy 电湿法冶金
electrojet 电喷流,电集流
electrokinetics 动电学,电动力学
electrokinetograph 动电计
electrologging 电测程
electroluminescence (EL) 电致发光
electrolysis 电解
electrolysis tank 电解槽
electrolyte 电解液,电解质
electrolytic analysis 电解分析
electrolytic bath 电解池
electrolytic capacitor 电解电容器
electrolytic deposition 电解沉积
electrolytic resistance 电解电阻
electrolyzer 电解装置,电解器;电解槽,电解池
electromachining 电加工
electromagnet 电磁铁,电磁体
electromagnetic amplifying lens (EAL) 电磁放大透镜
electromagnetic compatibility (EMC) 电磁兼容性

electromagnetic coil 电磁线圈
electromagnetic contactor 电磁接触器
electromagnetic field (EMF) 电磁场
electromagnetic forming 电磁成形
electromagnetic induction 电磁感应
electromagnetic interference (EMI) 电磁干扰
electromagnetic radiation (EMR) 电磁辐射
electromagnetic relay 电磁继电器
electromagnetics 电磁学
electromagnetic torque 电磁转矩
electromagnetic volume (EMV) 电磁量
electromagnetism 电磁学;电磁
electromechanical analogy 机电模拟
electromechanical coupling 机电耦合
electromechanical device 机电器件
electromechanical relay 机电继电器
electromechanical research (EMR) 电机械研究
electromechanical transducer 机电换能器
electromechanics 机电学
electromer 电子异体体,电子异构物
electromerism 气体电离过程;电子异构
electromerization 电子异构作用
electrometallization 电喷镀金属
electrometallurgy 电冶金学,电冶金法
electrometer 静电计
electrometrics 测电学
electrometry 测电术,测电法
electromicrometer 电子测微计
electromicrometry 电子测微法

electromotance 电动势
electromotive force (EMF) 电动势
electron 电子
electron accelerator 电子加速器
electron admittance 电子导纳
electron affinity 电子亲和势
electron beam 电子束
electron beam exposure system 电子束曝光系统
electron beam generator 电子束发生器
electron beam patterning 电子束成像
electron beam recorder (EBR) 电子束记录仪
electron beam scanning system (EBS system) 电子束扫描系统
electron capture 电子俘获
electron conduction 电子导电,电子传导
electron coupling 电子耦合
electron coupling control (ECC) 电子耦合控制
electron diffraction 电子衍射
electron drift 电子漂移
electronegative gas 电负性气体
electronic AC switch 交流电子开关
electronic automatic exchange (EAX) 电子自动交换机
electronic bionics 电子仿生学
electronic circuit analysis program (ECAP) 电子电路分析程序
electronic conductivity 电子电导率
electronic countermeasures (ECM) 电子干扰,电子对抗
electronic DC switch 直流电子开关
electronic generator 电子振荡器
electronic governor 电子稳速器
electronic guidance (EG) 电子制导
electron ionization 电子电离
electron optics 电子光学

electron pair 电子对
electron transition 电子跃迁
electron tube 电子管,真空管
electron vacancy 电子空位
electrooptics 电光学
electroosmosis 电渗(现象)
electrophore 起电盘
electrophoresis 电泳
electrophoretogram 电泳图
electrophorus 起电盘
electrophotometer 光电光度计
electrophysics 电物理学
electrophysiology 电生理学
electroplate 电镀;电镀物
electropneumatic brake 电空制动器
electropneumatic contactor 电气气动接触器
electropolishing 电抛光
electropositivity 正电性,阳电性
electroprobe 电测针,试电笔
electropsychrometer 电(测)湿度计
electropult 电气发射器
electropyrometer 电测高温计
electroquartz 电造石英
electrorefining 电精制,电解精炼
electroreflectance 电反射比,电反射率
electroresponse 电响应
electroretinogram (ERG) 视网膜电图
electroretinograph 视网膜电图描记器
electroretinography 视网膜电描记术,视网膜电图学
electroscope 验电器
electroscopy 验电法,气体电离检定法
electrosemaphore 电信号机
electroslag 电渣,电炉渣
electroslag welder 电渣焊机
electrosol 电溶胶
electrosorption 电吸附
electrostatic accelerator 静电加速器

electrostatic attraction 静电吸引
electrostatic deflection 静电偏转
electrostatic field 静电场
electrostatic induction 静电感应
electrostatic precipitator (ESP) 静电除尘器
electrostatics 静电学
electrostatic screening 静电屏蔽
electrostriction 电致伸缩,电缩作用
electrosynthesis 电合成
electrotaxis 趋电性
electrotechnics 电工
electrotechnology 电工学
electrothermal 电(致)热的
electrothermics 电热学
electrothermometer 电测温度计
electrothermy 电热学
electrotome 电刀;自动切断器
electrotropism 向电性,趋电性,应电性
electrotyping 电铸
electrovibrator 电振动器
electroviscosity 电黏滞性
electrowelding 电焊
electrowinning 电解提取
Elema 硅碳棒
element 元素;单元;元件
element antenna 振子天线
elevation 立面图;标高,高程;仰角
ELF (extremely low frequency) 极低频
EMC (electromagnetic compatibility) 电磁兼容性
emergency lighting 应急照明
emergency operating mode 事故运行方式
emergency power supply (EPS) 事故备用电源
EMF[1] (electromagnetic field) 电磁场
EMF[2] (electromotive force) 电动势
EMI (electromagnetic interference) 电磁干扰
emission 发射
emission current 发射电流
emission frequency 发射频率
emitter follower 射极输出器,射极跟随器
empire cloth 绝缘布
empire silk 绝缘丝
empty band 空带
empty level 空能级
empty running 空载运行
EMR[1] (electromagnetic radiation) 电磁辐射
EMR[2] (electromechanical research) 电机械研究
emulsifier 乳化剂;乳化器
emulsion 乳液;感光乳胶
emulsion plate 乳胶板
EMV (electromagnetic volume) 电磁量
enamel insulation 漆包绝缘
enclosed busbar 封闭母线
end bracket 端盖
end fitting 端头配件
energetics 能量学
energized condition 激励状态
energized position 激励位置,励磁位置
energizing 通电
energizing apparatus 励磁设备,激励设备
energizing circuit 励磁电路
energizing frequency 激励频率
energizing voltage 激励电压
engagement 啮合
EP[1] (earth plate) 接地板
EP[2] (electric propulsion) 电(力)推进
epicycloid 外摆线
epifilm 外延膜
epilayer 外延层
EPS[1] (electric power storage) 蓄电池
EPS[2] (electric power supply) 电力电源
EPS[3] (emergency power supply) 事故备用电源

Epstein frame 爱普斯坦方圈
equalization 均衡
equalization circuit 平衡电路
equalizer 均压线；均衡器
equalizing charge 均衡充电
equalizing ring 均压环
ER (earth return) 大地回路，接地回线
ERG (electroretinogram) 视网膜电图
ERW (electrical resistance welding) 电阻焊
ESP (electrostatic precipitator) 静电除尘器
event tree 事件树
exchanger 交换器，交换机；交换剂
excitation 励磁；激励；激发；激振
excitation response 励磁响应
excitation suppression 灭磁
excitation transformer 励磁变压器
excitation winding 励磁绕组
excited atom 受激原子
excited complex 受激络合物
exciter 激振器；励磁机；激励器

exciter response 励磁机响应
excitron 激励管；励弧管
exoergicity 放能度，释能度
exoergic reaction 放能反应，释能反应
extender 延长器
external insulation (EI) 外(部)绝缘
external overvoltage 外过电压
extra-high pressure discharge 超高压放电
extra-high tension (EHT) 超高压，极高压
extra-high vacuum 超高真空
extra-high voltage (EHV) 超高压，极高压
extremely high frequency (EHF) 极高频
extremely low frequency (ELF) 极低频
extruded insulation 挤包绝缘
eye bolt 活节螺栓，有眼螺栓
eye hook 眼钩
eye nut 有眼螺母
EZ (electrical zero) 电零，电零点

face shield （防护）面罩
facing 镀层,涂料
factorial experiment 析因实验
factor of earthing 接地因数
factory serial number 工厂序列号
fade-out （信号）衰减
fader 光量控制器
fail-safe 故障安全,保安性
fail-safe brake 安全制动器
fail-safe interlock 失效保护安全联锁,故障安全联锁（装置）
fail-soft mode 故障弱化方式
failure analysis 失效分析
failure cause 失效原因,故障原因
failure criticality analysis 临界失效分析,临界故障分析
failure diagnosis 故障诊断
failure load 破坏荷载,破坏负载
failure mode 失效模式
failure probability distribution 失效概率分布
failure rate 失效率
fair drawing 清绘
fairing 整流罩,整流片,减阻装置
fairlead 导缆器,导索器
fall time 下降时间;衰减时间
false firing 误通
false triggering 误触发
FAMOS memory FAMOS存储器,浮栅雪崩注入MOS（金属氧化物半导体）存储器
fan-cooled machine 风冷电机
fan-out 输出端数;扇出
farad 法,法拉（电容单位）

Faraday's law (of induction) 法拉第定律,电磁感应定律
far-field diffraction pattern 远场衍射图样
fascia 控制面板;（车辆）仪表盘
fast-acting fuse 快速熔断器
fast neutron flux 快中子通量
fast neutron reactor 快中子反应堆
fast neutron spectrum 快中子能谱
fast relay 高速继电器
fast scram 停堆
fatigue 疲劳
fatigue failure 疲劳破裂;疲劳失效
fatigue fracture 疲劳断裂
fatigue fracture test 疲劳断裂试验
fault 故障;层错
fault current 故障电流,短路电流
fault detection 故障探测,故障检测
fault handling 故障处理
fault impedance 故障阻抗,短路阻抗
fault initiating switch 快速接地（保护）开关
fault loop 故障回路
fault masking 故障屏蔽
fault phase 故障相
fault point 故障点
fault potential 故障电位
fault protective device 故障（电压）保护设备
fault section 故障区段
fault tolerance 容错

fault withstandability 耐故障能力, 故障承受能力
FDDI (fibre distributed data interface) 光纤分布式数据接口
feasibility 可行性, 可实现性
feeble field 弱场
feed 供电, 供电; 进料; 进刀
feedback 反馈
feedback control 反馈控制, 闭环控制
feedback element 反馈环节
feed drive 进给驱动; 进刀驱动
feeder 馈电线, 供电线, 馈路; 进刀机构; 供给装置
feeder breaker 馈电线断路器
feeder cable 馈电电缆
feeding point 供电点, 供电位置
feed motion 进给(运动)
feed per revolution 每转走刀量
feed rate 进给量
feed-through capacitor 穿心电容器
feeler 探针; 塞尺, 厚薄规
female connector 内孔连接器, 内螺纹连接器
female contact 阴接
female nipple 内接螺母
fender 碰垫, 舷材; 防擦物
Ferraris meter 费拉里斯电表
Ferraris motor 费拉里斯电动机
ferrite 铁氧体
ferroelectricity 铁电性
ferromagnetic oscillograph 铁磁录波器
fibre axis 光纤轴
fibre buffer 光纤缓冲层
fibre bundle 光纤束
fibre bundle jacket 光纤束护套
fibre coupler 光纤耦合器
fibre distributed data interface (FDDI) 光纤分布式数据接口
fibreglass-reinforced plastic (FRP) 玻璃纤维增强塑料
fibre joint 光纤接头
fibre-optic bus 光纤总线
fiducial value 基准值

field 场; 域; 字段
field coil 磁场线圈
field structure 场结构
filament 灯丝; 丝极
filler strip 垫片, 垫条
fillet 圆角
fillet weld 角焊缝
fill mode 填充方式
film boiling 膜(态)沸腾
film capacitor 薄膜电容器
filter 过滤器; 滤波器
filter capacitor 滤波电容器
filter cartridge 滤芯
filtration efficiency 过滤效率
fin 散热片, 肋片
final circuit 终接电路
final inspection 最终检查, 最终检验, 终检
fine adjustment 细调, 精密调整
fine feed 精细进给; 小步走刀
fine positioning 精定位
finish 精整度, 光洁度; 精加工
finish mark 加工符号; 光洁度符号
finite impulse response (FIR) 有限冲激响应
finite planing 精刨
FIR (finite impulse response) 有限冲激响应
fire point 燃点
firing 开通; 着火
firing failure 失通
firing stabilizer 稳燃器
firing valve 开通阀
firing voltage 着火电压
firm energy 保证电能
first base point 第一基点
first critical speed of rotation 第一临界转速
first-grade sheet 甲级板材
fishing wire 牵引线
fitter 装配工, 钳工
fitting 安装; 金具; 拟合
fit tolerance 配合公差
flame arc lamp 火焰弧光灯
flame arrester 阻火器

flameproof electrical apparatus 隔爆型电气设备
flame-retarding construction 阻燃结构
flat formation 平面敷设
Fleming's right-hand rule 弗莱明右手定则
flexible connection 软联结
flexible graphite 柔性石墨
flexible surface heater 挠性表面加热器
floating charge 浮充电
floodlight 投光灯;投光照明
flue gas 烟气
flue gas denitrification 烟气脱氮
flue gas desulphurization 烟气脱硫
flue gas emission control 烟气排放控制
fluidized-bed coating 流化床涂敷
fluidized-bed combustion 流化床燃烧
flux 通量;熔剂,焊剂
flywheel diode 飞轮二极管
foaming 泡沫共腾
focusing coil 聚焦线圈
focusing electrode 聚焦电极
fog lamp 雾灯
foil coil 箔式线圈
forbidden band 禁带
forced cooling 强迫冷却
form factor 波形因数;形状因数
forward breakdown 正向击穿
foundation cleaning 清基
foundation grouting 基础灌浆
foundation pit 基坑
Fourier series 傅立叶级数
fractional slot winding 分数槽绕组

fracture toughness 断裂韧性
frame-mounted motor 架承式电动机
free-wheeling arm 续流臂
frequency domain 频域
frequency relay 频率继电器
frequency-sensitive rheostat 频敏变阻器
FRP (fibreglass-reinforced plastic) 玻璃纤维增强塑料
fuel battery 燃料电池组
fuel blending 燃料配用
fuel cell 燃料电池
fuel measurement 燃料计量
fuel pellet 燃料芯块
fuel swelling 燃料肿胀
full load 满载
full-pitch winding 整距绕组
full-power tapping 满容量分接
full-wave arrangement 全波结构
fully controllable connection 全控联结
furnace 炉膛,炉胆
furnace arch 折焰角
furnace enclosure design pressure 炉膛设计压力
furnace explosion protection 炉膛防爆保护
furnace implosion 炉膛内爆
furnace transformer 电炉变压器
fuse 熔断器,保险丝
fused capacitor 带熔断器的电容器
fused short-circuit current 熔断短路电流
fuse element 熔件
fuse-link 熔断体
fusion frequency 停闪频率

Gg

gain 增益
gain adjustment 增益调节
gain coefficient 增益系数
gain constant 增益常数
gain control 增益控制
gain controller 增益控制器
gain crossover 增益交越
gain crossover frequency 增益交越频率
gain error 增益误差
gain flatness 增益均匀性
gain margin 增益裕量,增益边限
gain ripple 增益波动
gain slope 增益斜率
gallium-arsenide diode 砷化镓二极管
gallop 驰振
galvanic coupling 电流耦合
galvanic voltage 电流电压
galvanization 电镀,镀锌
galvanizing test 镀锌试验
galvanometer 检流计,灵敏电流计
galvanoplasty 电镀;电铸,电铸术
game theory 博弈论,对策论
gamma distribution 伽马分布
gammagraph 伽马射线照片
gamma-radiography 伽马射线照相术
gang control switch 联动控制开关
gang programmer 多重编程器
gang switch 同轴开关,联动开关
gap 间距,间隙
gap gauge 间隙规
gap loss 缝隙损耗
gas and vapour proof machine 气密与汽密型电机

gas-insulated bushing 气体绝缘套管
gas-insulated circuit 气体绝缘电路
gasket 垫圈,密封圈
gas leakage test 漏气试验,气密性试验
gas turbine set 燃气轮(发电)机组
gas welding 气焊
gate 门;闸门
gate circuit 选通电路,门电路
gate control 选通控制,门极控制,栅极控制
gated voltage 门极电压
gate electrode 栅极,门极
gate signal 门信号
gate terminal 栅极端子;栅极引出线
gateway 网关,信关
general-purpose branch circuit 通用分流电路,通用分路
general-purpose field communication system 通用现场通信系统
general-purpose fuse 通用熔断器
general-purpose interface bus (GPIB) 通用接口总线
general-purpose switch 通用(负荷)开关
general reset 总复位
general-service tungsten filament lamp 通用钨丝灯泡
general symbol 一般符号,通用符号
general tolerance 一般公差
generating capacity 发电容量,发电能力

generating gear 产形齿轮
generating plant 发电厂
generating set 发电机组,发电设备
generating station 发电厂
generator 发电机;发生器
generator bus 发电机母线,发电机引线
generator lead 发电机引线
generator output 发电机(功率)输出
generator pit 发电机机坑
generator rating 发电机额定值,发电机功率定额
generator rotor 发电机转子
generator terminal 发电机端子
generator transformer 发电机变压器
geometric distortion 几何畸变,几何失真
gland 密封套
glitch 假信号;突然失灵
goniometer 量角器
GPIB (general-purpose interface bus) 通用接口总线
graphitized carbon 石墨化碳
grid 栅极;(高压)输电网;网格栅;网格;骨架
grid bias 栅偏压
grid-controlled arc discharge tube 栅控弧光放电管
grid coordinate 网格坐标
grid coordinate system 网格坐标系
grid current 栅极电流
grid driving power 栅极激励功率
grid driving voltage 栅极激励电压
grid factor 栅格因素
grid input power 栅格输入功率
grid input voltage 栅格输入电压
grid point 坐标网点,网格点
grid pulse 栅极脉冲
grid-type earth electrode 网格式接地(电)极,网格接地体
gross baud rate 总波特率

gross generation 总发电量
gross head 总落差,总水头
gross installed capacity 总装机容量,总安装功率
gross output 总功率,总输出
gross thermal efficiency 总热效率
ground 地线,接地装置;接地
ground bus 接地母线,总地线,接地汇流排
ground clamp 接地夹,地线夹
ground clearance 离地净高,对地净空距离
ground conductor 地连接线,接地导体
ground connector 接地插头,接地接头
ground-coupled interference 地面耦合干扰
ground current 泄地电流,大地电流
ground-current equalizer 接地电流均衡器
ground-current limiter 接地电流限制器
ground detector 接地检测器,接地探测器
ground differential protection 接地差动保护
grounded circuit 接地回路
grounded neutral 接地中性点
grounded neutral system 中性接地制
ground fault 接地故障
ground-fault arc 接地故障电弧
ground-fault circuit interrupter 接地故障电路中断器
ground-fault circuit protection 接地故障电路保护
ground-fault compensation 接地故障补偿
ground-fault current 接地故障电流,对地短路电流
ground-fault detector 接地故障探测器
ground-fault location 接地故障位置;接地短路测定

ground-fault loop impedance 接地故障回路阻抗
ground-fault monitor 接地故障监视器
ground-fault neutralizer 接地故障中和器
ground-fault protection 接地故障保护
ground-fault reactor 接地故障电抗器
ground-fault relay 接地故障继电器
ground-fault test 接地故障试验
ground flash 落地雷
ground flash density 落地雷密度
ground indicator 接地指示器
grounding bar 接地线棒
grounding blade 接地刀
grounding cable 接地电缆
grounding electrode 接地电极
grounding jumper 接地跳线
grounding outlet 接地出线
grounding pad 接地板,接地垫
grounding switch 接地开关
grounding-type plug 接地型插头,带保护接地触点的插头
grounding-type receptacle 接地型插座,带保护接地触点的插座

ground insulation 对地绝缘,接地绝缘
ground joint 磨口接头;接地接头
ground leakage detection 漏地探测,对地短路探测
ground potential 大地电位,接地电压
ground resistance 接地电阻,大地电阻
ground return 接地返回电路
ground signal panel 地面信号板
ground wire 接地线,地线,避雷线
group delay 群时延
growing by pulling 拉制生长
growing by zone melting 区熔生长
growler 短路线圈测试仪
guard 保护装置,保护罩
guard band 防护频带
guarded input 保护输入
guard ring 保护环
guidance force 导向力
guidance system 制导系统
gunite 喷射混凝土
gutter 沟,槽
guyed V tower 拉线V型塔

Hh

half-bridge 半电桥
half-coiled winding 半圈式绕组
half-controlled bridge 半控制电桥
half-duplex interface 半双工接口
half-period 半周期,半衰期
half-wave rectifier 半波整流器
Hall effect 霍尔效应
Hall-effect device 霍尔效应器件
Hall-effect magnetometer 霍尔效应磁强计
Hall-effect pickup 霍尔效应探头
Hall-effect sensor 霍尔效应传感器
Hall-effect switch 霍尔效应开关
Hall generator 霍尔发生器
Hall mobility 霍尔迁移率
Hall modulator 霍尔调制器
Hall plate 霍尔板
Hall probe 霍尔探头
Hall voltage 霍尔电压
hand reset 手动复位,人工重调
hardware failure 硬件故障
harmonic 谐波
harmonic absorber 谐波吸收器
harmonic analysis 谐波分析
harmonic analyzer 谐波分析仪
harmonic compensation 谐波补偿
harmonic component 谐波分量
harmonic detector 谐波指示器,谐波检波器
harmonic filter 谐波滤波器,去谐滤波器
harmonic force wave 谐和力波
harmonic generator 谐波发生器
harmonic resonance 谐波谐振

harmonic restraint 谐波抑制,谐波稳定
harmonic test 谐波试验
HAZ (heat affected zone) 热影响区
hazard potential 危险电位
HBES (home and building electronic system) 住宅与楼宇电子系统
headed brush 带接头电刷
headend 终端站
header 联箱,集箱
heat affected zone (HAZ) 热影响区
heat balance 热平衡
heat capacity 热容(量)
heat consumption 热耗量
heat exchange coefficient 换热系数
heat exchanger 换热器
heat flux 热流通量
heating cable unit 加热电缆单元
heating capacity 制热量
heating circuit 加热电路
heating conductor 加热导体
heating resistor 加热电阻体
heat pump 热泵
heat recovery boiler 余热锅炉
heat run 耐热试验,热运行,持续短路试验
heat shield 隔热屏
heat sink 散热件;热汇
heat source 热源
heat transfer coefficient 传热系数
Heaviside effect 赫维赛德效应,集肤效应
heavy current 强电流,大电流

heavy-current bus 大电流汇流排,大电流母线
heavy-current connector 大电流连接器
heavy-current engineering 强电工程
heavy-duty circuit-breaker 大功率断路器
heavy electrical engineering 强电工程
HES application protocol (home electronic system application protocol) 家用电子系统应用协议
hexa-phase circuit 六相电路
high-breaking-capacity fuse 大切断功率熔断器,高分断能力熔断器
high-conductivity aluminium 高电导率铝,电工铝
high-conductivity copper 高电导率铜
high-conductivity polymer 高电导率聚合物
high-current-density region 电流密集区
high-current impulse 大电流脉冲
higher-level automation system 高层自动化系统,上位自动化系统
higher-level computer 上位计算机
higher-level control loop 上位控制回路
higher-level system 上位系统
higher-order delay element 高阶延迟元件
higher-order time delay 高次时延
highest voltage for equipment 设备最高电压,最高设备电压
highest voltage of a system 系统最高电压
high-field superconductor 高磁场超导体
high-frequency capacitance 高频电容
high frequency-changer set 高变频机组
high-frequency distortion test 高频失真试验
high-frequency welding 高频焊接
high-impedance differential protection 高阻抗差动保护
high-impedance fault 高阻抗故障
high-impedance fault to ground 对地高阻抗故障
high-interrupting-capacity fuse 大切断功率熔断器,高分断能力熔断器
high-level input current 高电平输入电流
high-level language 高级语言
high-level signal 高电平信号
high-level voltage 高电平电压
high-load-factor consumer 高负荷率用户
high-load-factor tariff 高负荷率电价
high-load hours 高负荷小时
high potential 高电位,高电势
high-power amplifier 大功率放大器
high-power motor 大功率电动机
high-power synthetic circuit 大功率综合电路
high-precision meter 高精度测量仪表
high-pressure discharge lamp 高压放电灯
high-rating transformer 大功率变压器
high-resistance fault to earth 对地高电阻故障
high-resolution image spectrometer 高分辨率成像光谱仪
high-response-rate voltage regulator 高响应速率调压器,高速反应调压器
high-sensitivity amplifier 高灵敏度放大器

high-speed air-blast breaker 压缩空气高速断路器,高速气吹式断路器
high-speed air magnetic breaker 高速磁吹式断路器
high-speed automatic reclosing 快速自动重合闸
high-voltage delay time 高压延时
high-voltage direct current (HVDC) 高压直流
high-voltage distribution 高压配电
high-voltage divider 高压分压器
high-voltage endurance 高电压耐用性
high-voltage pickup 高压传感器
high-voltage power supply 高压供电
high-voltage regulating transformer 高压调压变压器
high-voltage switch 高压开关
high-voltage terminal 高压端子,高压接头
high-voltage test 高压试验,耐压试验
high-voltage test technique 高压试验技术
high-voltage vacuum contactor 高压真空接触器
high-voltage winding 高压绕组
hob 滚刀,滚铣
hoist 起重机,升降机;启闭机
hoisting gear 提升机构,起重装置
hoisting tackle 提升滑车,起重滑车
hoistway 提升间,井道
holding value 持平值
hold-off interval 关断间隔
hold time 维持时间
hole diameter 孔径
hole semiconductor 空穴型半导体,P型半导体
hole storage effect 空穴存储效应
hole-type conductivity 空穴导电性;空穴电导率
holiday 漏点,漏斑
hollow insulator 空心绝缘子
hollow-shaft motor 空心轴电动机
hollow-shaft-type tachogenerator 空心轴型测速发电机
home and building electronic system (HBES) 住宅与楼宇电子系统
home electronic system application protocol (HES application protocol) 家用电子系统应用协议
home position 初始位置,原位
honeycomb radiator 蜂房式散热器
Hooke's coupling 万向接头,十字接头,万向联轴器
hook-on ammeter 钳形电流表
hook-up 连接;电路耦合;线路图,接线图
horizontal floor wiring cable 水平楼面接线电缆,分层接线电缆
horizontal frequency 水平扫描频率,行频
horizontal machine 卧式电机
horizontal parity check 水平奇偶校验
horizontal resolution 水平分辨率,水平清晰度
horsepower 马力
horsepower per machine volume 单位电机体积功率
horseshoe magnet 马蹄形磁铁
hose-proof machine 防喷型电机
host computer 主计算机
hot insertion 带电插入
hot restart 热再启动
hot standby 热备用
hot start-up 热态启动
hot unplugging 带电拔除插头
hot wire 热线;带电电线
house generator 厂用发电机
household (electrical) appliance 家用电器

housekeeping data 内务处理数据
house transformer 厂用变压器，自用变压器
housing 机壳
H-type cable H 型电缆
hub 轮毂；集线器
hum 交流声，低频噪音
human error 人为差错，人为误差
human factors engineering 人因工程学
human-machine interface 人机接口
hunting 追逐；连续自激振荡
HVDC (high-voltage direct current) 高压直流
HVDC converter transformer 高压直流换流变压器
HVDC substation 高压直流变电站
HVDCT (HVDC transmission) 高压直流输电
HVDC transformer 高压直流变压器
H-vector H 矢量
hybrid 混合；混合子
hybrid adaptor 混合式转接器，混合式适配器
hybrid circuit 混合电路
hybrid configuration 混合配置
hybrid connector 混合式连接器
hybrid data acquisition system 混合数据采集系统
hybrid drive 混合(动力)驱动
hybrid module 混合模块
hybrid multimeter 混合多用表，混合万用表
hybrid system 混合系统
hybrid UPS power switch 混合式不间断电源开关
hydraulic turbine 水轮机，液力涡轮机，水力涡轮机
hydrochemistry 水化学
hydroelectric set 水轮发电机组
hydrogen-cooled machine 氢冷电机
hydrogenerator 水轮发电机
hydropower plant 水电厂
hydropower station 水电站，水力发电站
hydrostatic pump 静液压泵
hysteresis loss 磁滞损耗

IC (integrated circuit) 集成电路
identification 辨识,识别
identification beacon 识别信标
identification code for converter connections 换流器电路识别码
identifier 标识符
idle capacity 空闲容量
idle circuit 空载电路
idle interval 不导通间隔
idler 空转轮,惰轮;导向轮,紧带轮
idle speed actuator 空转速度执行器
idle state 空闲状态
idle stroke 空行程
idle time 停机时间;空转时间,空载时间
idling 空转
igniter 点火器;点火极;点火剂;电弧发生器
ignition circuit 点火电路,引燃电路
ignition current 引燃电流,点火电流
ignition interaction 点火相互作用,预电离相互作用
ignitor 点火器;点火极;点火剂;电弧发生器
ignitron 引燃管,点火管
IIR (infinite impulse response) 无限冲激响应,无限脉冲响应
Ilgner generator set 伊尔格纳发电机组
Ilgner system 伊尔格纳系统
illegal operation 非法操作
illuminated mimic diagram 发光模拟图

illuminated switch 发光开关
image converter tube 变像管
image enhancement 图像增强
image register 图像寄存器
image sensor 图像传感器
imaging system 成像系统
immunity level 抗扰度电平
immunity to inference 抗干扰能力,抗扰度
impact moulding 冲击成型
impact notch test 缺口冲击试验
impact strength 冲击强度,撞击强度
impact stress 冲击应力
impact test 冲击试验
impact testing machine 冲击试验设备,冲击试验机
impedance 阻抗
impedance balance relay 阻抗平衡继电器
impedance converter 阻抗变换器
impedance coupling 阻抗耦合
impedance drop 阻抗压降
impedance-earthed system 阻抗接地系统
impedance earthing 阻抗接地
impedance loss 阻抗损耗
impedance matrix 阻抗矩阵
impedance ratio 阻抗比
impedance relay 阻抗继电器
impedance starter 阻抗起动器
impedance voltage at rated current 额定电流下的阻抗电压
impeller 叶轮,转子,波轮
impervious machine 防水机械,密封型电机
imported energy 输入能量

impress 外加(电流或电压)
impressed current 外加电流
impressed-current anode 外加电流阳极
impressed-current protection 外加电流保护
impressed voltage 外加电压
impression time 后效时间
improver 改进剂
impulse alternating current 脉冲交流电
impulse breakdown 脉冲击穿,冲击击穿
impulse breakdown strength 脉冲击穿强度,冲击击穿强度
impulse capacitance 脉冲电容
impulse chopped on the front 波头截断冲击波
impulse chopped on the tail 波尾截断冲击波
impulse crest voltage 最大冲击电压,冲击峰值电压
impulse current 冲击电流,脉冲电流
impulse-current limiter 冲击电流限制器
impulse flashover 脉冲击穿,冲击闪络
impulse flashover test 脉冲击穿试验,冲击闪络试验
impulse frequency 脉冲频率
impulse function 脉冲函数
impulse generator 脉冲发生器
impulse load 脉冲负荷,冲击负载,冲击负荷
impulse meter 脉冲计数器
impulse noise 脉冲噪声
impulse shape 脉冲波形
impulse sparkover test 冲击放电试验,脉冲击穿试验
impulse test 冲击试验
impulse testing station 冲击试验站
impulse torque 冲击力矩
impulse voltage 冲击电压
impulse withstand voltage 冲击耐受电压
impulsing mercury tube 脉动水银管
inactivity control 无通信监控
inadvertent contact 意外接触
inboard bearing 内装轴承
inboard rotor 内装转子
incandescent light 白炽灯
inching 点动,缓动,微动
inching control 微动控制,微控
inching mode 点动模式
inching speed 缓动速度,点动速度
incidence loss 冲角损失
incidental defect 伴随故障,偶然故障
incident illumination 反射照明
incipient break 初始破裂,初始断口
incipient crack 初始裂缝,初始裂纹
in-circuit emulator 内置仿真器,电路内部仿真器
incoming cable 输入电缆,供电电缆
incoming circuit-breaker 输入开关,供电开关,总开关
incoming cubicle 供电柜,配电间
incoming feeder 进线,进入馈线,输入馈线
incoming-feeder bay 进入馈线间隔,进线间隔
incoming inspection 输入检验
incoming overhead ground wire 进线避雷线
incoming panel 进线配电盘
incoming ring-feeder unit 环式馈线供电单元
incoming service aerial cable 房屋引入架空电缆
incoming service cable 房屋引入电缆
incoming unit 输入设备,供电设备
incomplete bridge connection 不完全桥式连接

incremental command 增量式指令
incremental compiler 增量编译器
incremental control 增量控制
incremental dimensioning 增量尺寸标注
incremental feed 增量式进给
incremental measuring method 增量测量方法
incremental mode 增量模式
incremental probability of failure 增量故障概率,增量失效概率
incremental programming 增量编程
incremental resistance 增量电阻
independent ballast 独立镇流器
independent circulating circuit component 独立循环回路部件,独立循环电路组件
independent current source 独立电流源
independent earthing 分离接地
indeterminate fault 非确定性故障,不确定故障
indexed access 索引存取
indexing hole 基准孔
indexing table 分度工作台,等分回转工作台
index method 指数法
index register 变址寄存器
indicated device 指示器
indicated value 指示值
indicating fuse 指示熔断器
indicating limit monitor 指示极值监视器
indicator board 指示器盘,显示板
indirect-acting instrument 间接作用式仪表
indirect cooled winding 间接冷却绕组
indirect-drive machine 间接驱动电机
indirect electric heating 间接电加热
indirect firing 间接点火
indirect lightning strike 感应雷击,间接雷击
indirectly air-cooled machine 间接空气冷却电机
indirect overcurrent release 间接过流脱扣器
individual mounting 单个装配
individual test 单独试验
indoor arrangement 室内布置
indoor bushing 户内套管
indoor earthing switch 室内接地开关
indoor external insulation 户内外绝缘
indoor substation 户内变电站
induced control voltage 感应控制电压
induced current 感应电流
induced draft 负压通风
induced draft fan 引风机
induced ignition 感应点火(器)
induced overvoltage withstand test 感应过电压耐受试验
induced voltage 感应电压
inductance 电感
inductance-capacitance coupling 电感电容耦合
inductance meter 电感表
inductance per unit length 单位长度电感量
induction 感应
induction coil 感应(线)圈
induction field 感应场
induction flux 感生通量,磁通(量)
induction frequency converter 感应变频机
induction furnace 感应炉
induction generator 感应发电机
induction meter 感应式电度表;感应计数器
induction starter 感应起动器,异步起动器
inductive ballast 感应镇流器
inductive breaking capacity 感应分断能力

inductive displacement transducer 电感式位移传感器
inductive load adjustment 电感性负载调节
inductive pickup 感应传感器
inductive potentiometer 感应电位器
inductive residual voltage 感应剩余电压
inductive shunt 感应分流器
inductivity 感应率;感应能力;介电常数
inductor 电感器;扼流圈
inductor-type synchronous motor 感应子同步电动机
industrial distribution equipment 工业配电装置
industrial robot data interface 工业机器人数据接口
inelastic impact 非弹性碰撞
inert gas 惰性气体
inert-gas seal 惰性气体密封
inert-gas-shielded welding 惰性气体保护焊
inertia 惯性;惯量
inertia compensation 惯性补偿
inertia constant 惯性常数
inertia torque 惯性力矩
infeed （横向）进给
infeed axis 进给轴
infeed motion （横向）进给运动,切深进给运动
infinite bus 无限大母线
infinite gain 无限增益
infinite impulse response (IIR) 无限冲激响应,无限脉冲响应
infinitely variable speed transmission 无级变速器,无级变速箱
influence by DC 直流感应
influence coefficient 影响系数
influence machine 静电发电机,感应起电机
influence of magnetic induction of external origin 外部磁感应影响
infrared absorption 红外吸收

infrared central locking system 红外中央控制闭锁系统
infrared controller 红外遥控器
infrared detector 红外探测器
infrared monochromator 红外单色仪
infrared radiation 红外辐射
infrared remote control system 红外遥控系统
infrared spectroscopy 红外光谱学,红外光谱法
infrared thermometer 红外温度计
inherent burden 固有负载
inherent delay 固有延迟
inherent error 固有误差
inherent feedback 固有反馈
inherent reliability 固有可靠性
inherent transient stability 固有瞬态稳定性
inherent weakness failure 本质失效,固有失效
inherited error 继承误差
inhibited oil 阻化油
inhibitor 抑制剂,阻化剂,阻聚剂,缓蚀剂
in-house line 用户线
initial capacitance 初电容
initial strain 初应变
initial stress 初(始)应力,起始应力
initial surge-voltage distribution 起始浪涌电压分布
initial symmetrical short-circuit current 对称短路电流初始值
initial torque 初始转矩
initial transient reactance drop 起始瞬态电抗降
initial transient recovery voltage (ITRV) 起始瞬态恢复电压
initial value of dead band 死区起始值
initial voltage 初始电压
initial watts 起始功率
initiation assignment 起动分配
initiator 发起站;引发剂

injected-beam magnetron 注入电子束磁控管
injected volume 注入剂量
injection current 注入电流
injection electroluminescence 注入电致发光
injection energy 注入能量
injection laser diode 注入式激光二极管
injection level 注入水平
injection locking range 注入同步范围
injection method 注入法
injection moulding 热压铸,注射成型,注塑
injection part 注压件,注塑件
injection pump 喷水泵,喷油泵
injection transformer 注入变压器
injection valve 喷射阀
inlet 进口,入口;插入物
inlet air duct 供风道
inlet pressure 进气压力,吸入压力
in-line package switch 成列直插封装开关
inner bearing ring 内轴承环
inner-cooled conductor 内部冷却导体,内部冷却导线
in-phase booster 同相升压机
in-phase control 同相控制
in-phase current 同相电流
in-phase null voltage 同相零位电压
in-plant power station 自用发电站
in-process inspection 工序间检验,加工过程检测
input 输入;入口;输入装置
input capacitance 输入电容
input circuit 输入电路
input energizing quantity 输入激励量
input immittance 输入导抗
input offset current 输入偏移电流,输入失调电流
input/output 输入输出

input/output level 输入输出电平
input/output pair 输入输出对
input prompt 输入提示
input regulation coefficient 输入调节系数
input resistance 输入电阻
input resolution 输入分辨率
input shaft 输入轴
input step 输入步长
input threshold voltage 输入阈电压
input transfer rate 输入速率
input triggering voltage 输入触发电压
input unit 输入单元
input variable 输入变量
input voltage range 输入电压(工作)范围
input winding 输入绕组
inrush current 涌流
inrush making current 接通涌流
inrush peak 接通电流峰值,冲击电流最大值
inrush restraint 涌流抑制
inrush suppressor circuit-breaker 冲击抑制电路断路器
inrush transient current 合闸瞬态电流,涌流
insensitivity 不灵敏性,不敏感性
inserted radius 插入半径
insertion and withdrawal force 插入拔出力
insertion current 插入电流
in-service inspection 在役检查
in-service period 承载时间
in-service test 运转试验
inside calipers 内卡钳,内径测量仪
inside contour 内部轮廓
inside diameter 内径
inside dimension 内部尺寸
inside width 净宽度,内宽
insignificant non-conformance 可忽略误差
in-situ balancing 现场平衡
in-situ concrete 现场浇注混凝土

in-situ maintenance 现场维修
inspection 检验,检查
inspection box 观察箱,检查箱
inspection by attributes 计数型检查
inspection by variables 计量型检查
inspection earthing 检修接地
inspection gauge 检验量规
inspection hole 观察孔
inspection lot 检查批
inspection report 检验报告
inspection sticker 检验标签
installation 安装,装配,铺设;(大型)设备
installation clearance 安装间距
installation cost 安装成本
installation under plaster 暗装布线
installation under the surface 暗装布线
installed capacity 装机容量,设备容量
installed life 安装后的使用寿命
instantaneous current 瞬时电流
instantaneous operation 瞬时动作
instantaneous power 瞬时功率
instantaneous protection 瞬时保护
instantaneous release 瞬时脱扣器
instantaneous trip 瞬时切断
instant of chopping 截断瞬间
instant-start lamp 快速起动灯,冷起动灯
instruction 指令,指示;操作说明
instruction code 指令码
instruction decoder 指令译码器
instruction set 指令集,指令系统
instrumentation amplifier 测量放大器
instrumentation and control 测控
instrument case 仪表盒,仪表箱
instrument security current 仪表安全电流
instrument security factor 仪表保安因数
instrument transformer 互感器
instrument with magnetic screen 磁屏蔽仪表
instrument with optical index 光标式仪表
insulance 绝缘电阻
insulant 绝缘体,绝缘材料
insulated cable 绝缘电缆
insulated clamp 绝缘夹钳
insulated coupling 绝缘连接;绝缘联轴节
insulated-shield cable system 屏蔽绝缘电缆系统
insulating ability 绝缘能力
insulating agent 绝缘介质
insulating cement 绝缘胶
insulating clearance 绝缘间隙
insulating coating 绝缘涂层
insulating property 绝缘性质
insulating tape 绝缘带,绝缘胶带
insulation 绝缘;隔离;绝缘材料
insulation breakdown 绝缘击穿
insulation class 绝缘等级
insulation cost 绝缘成本
insulator 绝缘体;绝缘子
insulator string 绝缘子串
insulator-type current transformer 绝缘子式电流互感器
integral control 积分控制
integrally fused circuit-breaker 带熔断器的断路器
integral switching device 整体式开关装置
integrated air/fuel system 集成空气/燃油系统
integrated circuit (IC) 集成电路
integrated-circuit relay protection system 集成电路继电保护系统
integrated computer-aided manufacturing 集成计算机辅助制造
integrated data network 综合数据网

integrated microcircuit 集成微电路
integrated optical circuit 集成光路
integrated service digital network (ISDN) 综合业务数字网
intelligent charger 智能充电器
intelligent field device 现场用智能设备
intended conditions of use 设定使用条件
intended life 设定(使用)寿命
intensity 强度
interaction 交互作用,相互作用
interaction field 互作用场
interaction gap 互作用隙
interactive mode 交互方式
interactive processing 交互式处理
intercell connector 连接条
interchange circuit 互换电路
interchange function 交换功能(数控加工中的刀具更换)
interchange power 交换功率,互换功率
intercoil insulation 线圈间绝缘
interconnected operation 互连运行,互连运转
interconnection 互连;连接器
interconnection mask 互接掩模
interconversion 互换,互变
interface 接口;界面;接通,连接
interface agent 接口主体
interface assignments 接口分配
interface bus 接口总线
interface effect 界面效应,边界效应
interface handler 接口管理器;界面管理器
interface management bus 接口管理总线
interface protocol 接口协议
interface reaction-rate constant 界面反应率常数
interference 干扰;干涉
interference effect 干扰效应
interference fading 干扰衰落

interference field strength 干扰场强
interference filter 干扰滤波器,干扰滤光片
interference source 干扰源
interference suppression 干扰抑制
interference suppression choke 干扰抑制扼流圈
interference suppression equipment 干扰抑制装置
interference suppression symbol 干扰抑制符号
interference voltage 干扰电压
interference voltage meter 干扰电压测量仪
interframe time fill 帧间时间填充
interim test 中间测试
interior lighting 室内照明
interlaced display 隔行交错显示
interlacing impedance voltage 交错阻抗电压
inter-lamination fault 层间短路
interlayer 中间层,夹层
interlayer connection 层间连接
interlayer insulation 层间绝缘
interlayer voltage 层间电压
interleaved coil 纠结线圈
interleaved double coil 纠结式双线圈
interleaved joint 交错结合点
interleaved phase windings 纠结式相绕组
interlinked flux 交链磁通
interlinked leakage flux 交链漏磁通
interlinking factor 交链系数
interlock bypass 联锁旁通
interlock bypass switch 联锁旁通开关
interlock cancelling 联锁解除
interlock deactivating key 解除联锁钥匙,分断钥匙
interlock deactivating means 解除联锁装置,分断装置
interlocked bus signal 相关总线信号

interlocked socket outlet 联锁插座
interlocked switched socket outlet 联锁可断开插座
interlocking circuit 联锁电路
interlocking device 联锁机构
interlocking electromagnet 联锁电磁铁
interlocking module 联锁模块
interlocking signal 联锁信号
interlocking switch group 联锁开关装置，联锁组合开关
intermediate cooling circuit 中间冷却回路，中间散热回路
intermediate electrode 中间电极
intermediate equipment 中间设备
intermediate form 中间形态
intermediate frequency amplifier 中频放大器
intermediate frequency transformer 中频变压器
intermediate gear box 中间齿轮箱
intermediate layer 中间层
intermediate measurement 中间测量
intermediate point 中间点，插点
intermediate support 中间支撑
intermediate switch 中间开关
intermediate system 过渡系统，转接系统
intermediate terminal 中间端子
intermediate voltage terminal 中间电压端子
intermediate voltage winding 中压绕组
intermediate yoke 中轭
intermittent arcing ground 间断性电弧接地
intermittent current 断续电流，间歇电流
intermittent duty 断续工作制，（电机）间歇运转
intermittent earth fault 间断接地故障
intermittent electric contact 电接触不良
intermittent failure 间歇失效
intermittent fault 断续故障，间歇故障
intermittent feed 断续进给
intermittent flow 断续流通
intermittent operation 间断运行
intermittent rating 间歇运转测量值；间歇运转测量；间歇运转功率
intermittent service test 断续运行试验
intermodulation 互调，相互调制
intermountability 安装互换性
internal arcing fault 内部电弧故障
internal brush drop 电刷内电压降
internal capacitance 内部电容
internal clock 内置时钟
internal combustion engine 内燃机
internal discharge 内部放电，内部释放
internal drift field 内部漂移场
internal efficiency 内效率
internal electric field 内建电场
internal equivalent voltage 内部等效电压
internal fault 内部故障
internal fault test 内部故障测试
internal immunity 内部抗干扰性
internal insulation 内绝缘
internal overpressure disconnector 内部过压断路器
internal overvoltage 内过电压，内部过压
internal photoelectric effect 内光电效应
internal residual voltage 内部剩余电压
internal resistance 内阻
internal state variable 中间状态变量
internal thread 内螺纹
internal viewing system 内部观测系统

internal wiring 内接线
international system of units 国际单位制
internetworking protocol 网际互连协议
interphase insulation 相间绝缘
interphase reactor 平衡电抗器,相间电抗器
interphase short circuit 相间短路
interphase transformer 相间变压器
interpole 中间极,换向极
interpole winding 中间极绕组,换向极绕组
interposing relay 中介继电器
interrupt arrangement 中断排列
interrupt chamber 灭弧室
interrupt confirmation 中断确认
interrupt controller 中断控制器
interrupt enable 允许中断
interrupter 灭弧室;中断器,分断器,断路器
interrupt function 中断功能
interrupt handler 中断处理器
interrupt handling system 中断处理系统
interruptible load 可中断负荷,可中断负载
interrupting capacity 断开容量,断路容量
interrupting current 断开电流,断路电流
interrupting device 断路装置
interrupting medium 断路媒质
interrupting rating 断续额定值
interrupting test 断路试验,断续试验
interrupting time 中断时间,断路时间
interrupt input 断续输入
interruption 中断,断开
interruption duration 停电时间,断电时间
interruption key 断路电键
interruption of power supply 断电,供电中断
interruptive ratio 切断比

interrupt level 中断等级
interrupt list 中断表
interrupt mask 中断屏蔽
interrupt overflow 中断溢出
interrupt priority (level) 中断优先级
interrupt processing 中断处理
interrupt request 中断请求
interrupt routine 中断程序
interrupt service routine (ISE) 中断服务程序
interrupt stack 中断堆栈
interrupt timer 中断计时器
interrupt vector 中断向量,中断矢量
interrupt vectoring 中断向量[矢量]控制
interrupt word overflow 中断字溢出
intersection 交;交面,交点
intersection box 十字接线盒
intersection cutter radius compensation 切割点铣刀半径轨迹校正
interstage coupling 级间耦合
interstage transformer 级间变压器,中间变压器
inter-stand short circuit 部分导体短路,绕组线段短路
intersystem fault 系统间故障
intersystem interference 系统间干扰
intertripping 联锁跳闸
intertripping underreach protection 联锁跳闸欠范围保护
interturn breakdown test 匝间击穿试验
interturn fault 匝间故障
interturn insulation 匝间绝缘
interturn short circuit 匝间短路
interturn test 匝间试验
interval 间隔;间隙;距离;区间
interval counter 间隔时间计数器
interval time-delay relay 延时间隔继电器

intervention by manual control 手控干预
interwinding fault 绕组间故障
interworking unit 交互工作单元
intra-plant system 厂内系统
intra-system interference 系统内部干扰
intrinsically safe circuit 本质安全电路
intrinsically safe electrical apparatus 本质安全型电气设备
intrinsic conduction 本征导电
intrinsic conductivity 本征导电率,本征导电率
intrinsic consumption 本身消耗
intrinsic damping 固有衰减
intrinsic electric strength 本征抗电强度
intrinsic error 固有误差
intrinsic induction 本征感应
intrinsic junction transistor 本征结型晶体管
intrinsic magnetic moment 固有磁距
intrinsic magnetic property 内禀磁性
intrinsic safety 本质安全(防爆)型,本安(防爆)型
intrinsic safety barrier 本质安全位垒
intrinsic semiconductor 本征半导体
intrinsic stand-off ratio 本征变位比
intruder alarm system 入侵报警系统
invalid reception 无效接收
invariable resistor 固定电阻器
inverse-acting controller 反转控制器
inverse-characteristic relay 逆转特性继电器
inverse current 逆电流,反向电流
inverse direction of operation 工作反向

inverse electrode current 反向电极电流
inverse excitation 反向激励,反向励磁
inversely magnetized 反向磁化的
inverse parallel connection 反向并联,逆并联
inverse reactance 反向电抗
inverse-speed motor 反速电动机
inverse time-delay operation 反时延动作
inverse time-delay overcurrent release 反时延过(电)流脱扣器
inverse-time feed rate 逆时进给率
inverse-time grading 逆时分级整定
inverse-time lag 逆时延时
inverse-time overcurrent protection 反时限过(电)流保护
inverse transformation 逆变换,反向变换
inversion 反向,倒转;倒反,反演
inversion density 反演密度
inversion efficiency 逆变效率
inversion factor 逆变因数
inversion layer 反型层
inversion of point patterns 点模式转换
inversion point 转换点
inverted machine 反转电机
inverted motor 反接电动机
inverted rotary converter 反向旋转变流机
inverted siphon 倒虹吸管
inverter 逆变器;反相器
inverter stability limit 逆变器稳定限度
inverter termination control 逆变器终止控制
inverter trigger set 逆变器触发装置
inverting amplifier 反相放大器
inverting negative impedance converter 反向负阻抗变换器

invocation 调用,启用
involute connection 渐开线连接
involute winding 渐开线型绕组
in-zone fault 区内故障;内部缺陷
ion 离子
ionic conduction 离子导电
ionic valve device 离子阀器件
ion implantation 离子注入
ion-implanted MOS circuit 离子注入 MOS 电路
ionization 电离,离子化
ionization detector 电离探测器
ionization discharge 电离放电
ionization extinction voltage 电离熄火电压
ionization inception voltage 电离起始电压
ionization phenomenon 电离现象
ionization probability 电离概率
ionization ratio 电离比
ionization smoke detector 离子烟雾探测器
ionization threshold 电离阈值
ionization tube 电离管
ionization voltage 电离电压
ionized cluster beam 离化团粒束
ionizing energy of acceptor 受主电离能
ionizing energy of donor 施主电离能
ionizing impurity 电离杂质
ionizing radiation 电离辐射
ion-selective electrode 离子选择电极
ion-selective field effect transistor 离子选择场效应晶体管
ion trap 离子阱
IR drop 电阻电压降
iron-core coil 铁芯线圈
iron-cored electrodynamic instrument 铁芯电动仪表
iron-cored ferrodynamic ratio meter 铁磁电动式比值计
iron-core reactor 铁芯电抗器
ironing width 熨平宽度
iron loss 铁耗
iron sheet 铁板
irradiance 辐照度
irradiation saturation current 辐照饱和电流
irregular transition 无规律转换
irrotational field 无旋场
ISDN (integrated service digital network) 综合业务数字网
ISE (interrupt service routine) 中断服务程序
isentropic exponent 等熵指数
island 岛(离散系统中的自动化单元)
island effect 小岛效应
island of automation 自动化孤岛
island of production 加工岛
isoacoustic curve 等声强曲线
isocandela diagram 等发光强度线,等坎德拉图
isochronous governor 等时调节器
isodromic governor 恒值调节器,匀速调节器
isodynamic governor 等力调节器
ISO fundamental tolerance series ISO 基准公差系列
isokeraunic level 等雷电级;年平均雷电日水平
isolated input 隔离输入
isolated lot 零星批量
isolated menu 独立菜单
isolated neural 不接地中线
isolated neural system 中性点绝缘系统
isolated output 隔离输出
isolated-phase bus 分相母线,隔离相母线
isolated position 零位;隔离位置,分离点
isolated power plant 独立电厂,孤立电厂
isolated supply 不接地电源
isolating blade 分离闸刀
isolating blade terminal 闸刀式隔离端子
isolating contact 分离触点

isolating distance 隔离段；绝缘段
isolating inductor 隔离电感器
isolating link 隔离开关
isolating measuring terminal 隔离测量端子
isolating neural terminal 隔离中性点端子
isolating plug connector 隔离插塞连接器
isolating point 分离点，隔离点
isolating submodule 隔离子模块
isolating switch 隔离开关
isolating terminal 隔离端子
isolating voltage transformer 隔离电压互感器
isolation 隔离；隔离度；绝缘
isolation amplifier 隔离放大器
isolation diffusion 隔离扩散
isolation from earth 不接地，对地绝缘
isolation from supply 切断电源
isolation of a unit 成套装置隔离，机组转入单机运行
isolation voltage 隔离电压
isolator 绝缘体；隔离器，隔离开关
isothermal controller 等温控制器
item designation 项目代号
iteration factor 迭代因子
iteration statement 迭代语句
iterative earthing 重复接地
iterative impedance 重复阻抗，累接阻抗
iterative network 累接网络
ITRV (initial transient recovery voltage) 起始瞬态恢复电压
IT system IT 系统

jabber 逾限(传输)
jabber control 逾限控制
jack 千斤顶;插孔,插口
jack bolt 起重螺栓,定位螺栓
jack distributor 插座配电盘
jacking device 升降机,升降装置,起重装置
jacking oil pump 顶轴油泵
jack panel 插孔板
jack screw 螺旋千斤顶,起重螺旋
jack-screw system 螺旋千斤顶系统
jack shaft 中间轴,传动轴
jackstay 支索
jerk limitation 冲击限制
jerk rate 冲击率
jet compressor 喷射压缩机
jig 夹具;钻模样板
jitter 抖动,(图像)跳动
job data 作业数据
job processing 作业处理
job queue 作业队列
job scheduling 作业调度
jog control 微动控制
jogging 微动

jogging speed 微动速度
Johnson noise 约翰逊热噪声,热噪声
joint 接头;接合;接缝
joint box 接线盒;电缆套管
joint welding 搭焊;接头焊接
Jordan bearing 推力套筒轴承
Josephson effect 约瑟夫森效应
Joule effect 焦耳效应
journal bearing 轴颈轴承
joystick 操纵杆
joystick transformer 调压变压器
jumper 跨接线,跳线
jumper assignment 跳线分布,桥接配置
jumper cable 跨接电缆,桥接电缆
jumper clamp 跳线线夹
jumper sag 跳线弛度
jump frequency 跃迁频率,跳变频率
jump operation 跳转操作
junction 结;连接;接头
junction box 接线箱,分线盒
junction transistor 结型晶体管
just value 适时值

K k

K (kelvin) 开,开尔文
KA (kiloampere) 千安(培)
Kaiser effect 凯泽效应
kaleidophone 示振器
kalimeter 碳酸定量器
kallirotron 负阻抗管,卡利罗管
Kalman cycle 卡尔曼周期
Kalman filter 卡尔曼滤波器
kaolin 高岭土,陶土
kaolinite 高岭石
Kaplan turbine 卡普兰式水轮机,轴流式水轮机
kapnometer 烟密度计
Kapp coefficient 卡氏系数
Kapp-Hopkinson test 卡普-霍普金森试验
karma foil 卡玛箔
Karman's vortex 卡门涡流
Karnaugh map 卡诺图
karst 喀斯特,岩溶
Kb (kilobyte) 千字节
K-band K 波段,K 频带
kbar (kilobar) 千巴
kc (kilocycle) 千周
kcal (kilocalorie) 千卡,大卡
K-capture K 俘获
K-carrier system K 型载波系统
K-conversion K 转换
K-display K 型显示,位移距离显示
keep-alive 保活,维弧
keeper 定位件,夹子
K-electron K 电子
Keller furnace 凯勒氏电弧炉
kelvin (K) 开,开尔文
Kelvin balance 开尔文秤
Kelvin bridge 开尔文电桥

Kelvin degree 开尔文温度,绝对温度
Kelvin double bridge 开尔文双电桥
Kelvin effect 开尔文效应,趋肤效应
Kelvin scale 绝对温标,开氏温标
KEM (kilocycle electromagnetics) 千周波电磁法
Kendall effect 肯德尔效应,假象效应
kenotron 高压整流二极管,二极真空整流管
Kerr cell 克尔盒
Kerr effect 克尔效应
kettle 深槽;汽锅
keV (kilo-electronvolt) 千电子伏(特)
kevatron 千电子伏(特)级加速器
key 键;电键,钥匙;栓,销;键控;键控法发报;用键固定
keybar 键棒
key bed 键槽
keyboard 键盘
keyboard processor 键盘处理器
keyboard switch 键盘开关
key coder 键盘编码器
keyer 键控器;定时器
key frame 关键帧
key generator 密钥生成器
keying circuit 键控电路
keying frequency 键控频率,电键频率
keying signal 键控信号
keying wave 键控信号波
keyless socket 无键插座
keypad 键区,小键盘

key pulse (KP) 键控脉冲
keypunch 键控穿孔;键控穿孔机
key relay 键控继电器
key seat 键槽
keyset 配电板;键盘
key shelf 键架
keystone distortion 梯形畸变,梯形失真
key switch 按键开关,键式开关
key tube 键控管
key wall 刺墙
keyway 键槽
K-frame structure K型架构
KG (kilogauss) 千高斯
kg (kilogram) 千克
KGV (knife gate valve) 刀闸阀
kHz (kilohertz) 千赫(兹)
kibbler 粉碎机
kick 冲,反冲力;突震
kickback 反冲,返程
kicker 喷射器;弹踢器
kicksort 脉冲振幅分析
kicksorter 脉冲振幅分析器
kick starter 反冲式起动器
kidney joint 挠性接头
killed line 断线
killed steel 镇静钢,全脱氧钢
killer 断路器;限制器
killer circuit 抑制电路
killer winding 灭磁绕组
kiloampere (KA) 千安(培)
kilobar (kbar) 千巴
kilobit 千位
kilobyte (Kb) 千字节
kilocalorie (kcal) 千卡,大卡
kilocurie 千居里
kilocycle (kc) 千周
kilocycle electromagnetics (KEM) 千周波电磁法
kilodyne 千达因
kilo-electronvolt (keV) 千电子伏(特)
kilogauss (KG) 千高斯
kilogram (kg) 千克
kilohertz (kHz) 千赫(兹)
kilojoule (kJ) 千焦(耳)
kilolambda 千微升
kiloline 千磁力线
kilolitre (kl) 千升
kilolumen 千流明
kilolux 千勒(克斯)
kilomega 千兆
kilomegabit 千兆位,十亿位
kilomegacycle 千兆周
kilo-oersted 千奥(斯特)
kiloroentgen 千伦琴
kilovar (kvar) 千乏
kilovar-hour 千乏时
kilovolt (kV) 千伏(特)
kilovoltage 千伏电压
kilovolt-ampere (KVA) 千伏安
kilowatt (kW) 千瓦
kilowatt-ampere (kWa) 千瓦安
kilowatt-hour (kWh) 千瓦时,度
kindling point 燃点
kinescope 显像管
kinetic current 动力电流
kinetic energy 动能
kinetic equilibrium 动态平衡,动力平衡
kinetic momentum 动量
kinetics 动理学,动力学
kingbolt 中心销,主螺栓
king post 吊杆柱,主梁柱
king-post antenna 主轴式天线
Kingsbury bearing 金斯伯里轴承
king valve 主阀,总阀
kink 扭折
kip 千磅
Kirchhoff's law 基尔霍夫定律
Kirchhoff's voltage law (KVL) 基尔霍夫电压定律
kJ (kilojoule) 千焦(耳)
kl (kilolitre) 千升
klirr 波形失真,非线性失真
klystron 速调管,速度调制管
klystron oscillator 速调管振荡器
knee 拐点;肘板;扶手弯头
knee bend 弯头
knee bracing 角撑,斜撑
knee frequency 拐点频率
knee-point voltage 拐点电压

knife contact 刀形触头
knife edge pointer 刃形指针
knife gate valve (KGV) 刀闸阀
knife switch 刀开关,闸刀开关
knob 旋钮
knocking 爆击
knocking combustion 爆燃
knock-inhibiting additive 抗爆剂
knockout 顶出器,顶脱件
knock pin 定位销
knock property 爆震性
knock rating 防爆率
knock test 爆震性试验
Knoop hardness number 努普硬度值
Knoop hardness penetrator 努普硬度压头
knot (波)节(海里/小时)
knuckle joint 万向接头
Knudsen's burette 克努森滴定管
Knudsen's pipette 克努森移液管
Kodak Standard perforation 柯达标准片孔
Kohlrausch bridge 柯尔劳希电桥
konimeter 测尘器
koniology 微尘学
konitest 计尘试验
KP (key pulse) 键控脉冲
kraft (paper) 牛皮纸
Krarup cable 连续加感电缆
K-series K系
K-shell K(电子)壳
kV (kilovolt) 千伏(特)
KVA (kilovolt-ampere) 千伏安
kvar (kilovar) 千乏
KVL (Kirchhoff's voltage law) 基尔霍夫电压定律
kW (kilowatt) 千瓦
kWa (kilowatt-ampere) 千瓦安
kWh (kilowatt-hour) 千瓦时,度
kymogram 记波图
kymograph 波形自记器,记波器
kymography 记波法

l¹ (length) 长度
l² (litre) 升
label coding 标号编码
label constant 标号常数
label data 标号数据
labelling 加标
lacing wire 拉筋
lacquer 漆,清漆,硝基漆
lacquer disk 蜡克盘
lacquer film 喷漆薄膜
lacquer layer 硝基漆层
ladder 梯
ladder circuit 梯形电路
ladder diagram 梯形图
ladder effect 梯形效应
ladder filter 梯式滤波器
ladder network 梯形网络
L aerial L形天线
lag 滞后
lag bolt 方头螺栓
lag compensation 滞后补偿
lagging current 滞后电流
lagging load 滞后负载
lagging phase 滞后相,弱相
lag module 滞后组件
lag network 滞后网络
Lagrange('s) equation 拉格朗日方程
Lagrangian method 拉格朗日法
lag theorem 延迟定理
lambert 朗伯
Lamb shift 兰姆移位
Lamb wave 兰姆波
lamella 薄片,薄板;片晶
lamellar field 非旋场
lamina 薄板,薄层
laminar deposition 层状淀积

laminar flow 层流
laminar plasma torch 层流等离子枪
laminate 层压板
laminated antenna 叠层天线
laminated core 叠片铁芯
laminated insulation 叠片绝缘
laminated magnet 叠片磁铁,积层磁铁
laminated tube 层压管
lamination 叠片;成层
lamination factor 分层系数
lamination insulation 片间绝缘
lamination varnish 叠片漆
lamp mount 灯芯
lamp tube 灯管
lance 喷枪
land 连接盘
landing-area floodlight 着陆区投光灯
Landau damping 朗道阻尼
land cable 地面电缆
lane 巷道
Langmuir frequency 朗缪尔频率,等离子体频率
L antenna L形天线
lap 研磨;研磨模
lap joint 搭接接头
Laplace-Gauss distribution 拉普拉斯-高斯分布
Laplace operator 拉普拉斯算子
Laplace plane 拉普拉斯平面
Laplace's equation 拉普拉斯方程
Laplace's field 拉普拉斯场
Laplace's law 拉普拉斯定律
lap-over 搭接
lapped insulation 绕包绝缘

lapped wire 绕包线
lapping compound 研磨剂
lapping hardness 研磨硬度
lap welding 搭接焊
lap winding 叠绕组
large and medium DC motor 大中型直流电动机
large electronic display (LED) 大电子显示器
large-scale integration (LSI) 大规模集成电路
Larmor frequency 拉莫尔频率
LAS (light-activated switch) 光启开关
LASCR (light-activated SCR) 光启可控硅整流器
lasecon 激光转换器
laser 激光;激光器
laser flash tube (LFT) 激光闪光管
laser image converter (LIC) 激光变像器
laser interference filter (LIF) 激光干扰滤波器
laser optical modulator (LOM) 激光光学调制器
laser welding 激光焊接
LASS (light-activated silicon switch) 光启硅开关
latch 锁存器
latched contactor 锁扣接触器
latched relay 闭锁继电器
latching circuit 锁住电路
latching current 擎住电流,闭锁电流
latching device 锁扣机构
latching electromagnet 闭锁电磁铁
latching relay (自)保持继电器
latency (time) 等待时间
latent electricity 束缚电
lateral deflection 横向偏转
lateral isolation 横向绝缘
lateral transistor 横向晶体管
lath 板条
latitude 纬度;纵距

latitude effect 纬度效应
lattice coil 蜂房式线圈
lattice constant 晶格常数
lattice defect 晶格缺陷
lattice distortion 晶格畸变
lattice matching 晶格匹配
lattice spacing 栅距
lattice tower 桁架式塔架
lattice vacancy 晶格空位
lattice vibration 晶格振动
Laue method 劳厄法
law of electromagnetic induction 电磁感应定律
law of electrostatic attraction 静电吸引定律
law of regulating action 调节动作定律
law of similarity 相似定律
Lawrence tube 劳伦斯管
layer coil 层式线圈
layered cable 层式电缆
layered mask 多层掩膜
layered medium 分层介质
layering 分层
layer insulation 层间绝缘
layer-to-layer signal transfer 层间信号串扰
layer winding 分层绕组
layout 布置;布局
layout chart 布局图
layout data 布置数据
layout plan (平面)布置图
lay ratio 扭绞系数
LB[1] (line busy) 占线
LB[2] (local battery) 本地电池
LC[1] (line concentrator) 集线器,线路集中器
LC[2] (link circuit) 链式电路
LC[3] (loaded cable) 加感电缆
L-cathode L型阴极,多孔隔板阴极
LCVM (log conversion voltmeter) 对数变换伏特计
LDA (line driving amplifier) 线路激励放大器
LDR (light-dependent resistor) 光敏电阻

lead¹ 引线,导线;超前
lead² 铅
lead bonding 引线键合
lead clad cable 铅包电缆
lead compensation 超前补偿
lead covered cable 铅包电缆
lead covering 铅皮
leader 先导;引线
leader cable 引线电缆
lead extrusion 压铅
lead-in 引入;引入线,输入端
leading current 超前电流
leading edge 前缘
leading-out grounding 外引式接地
leading-out terminal 引出端
leading phase 超前相,强相
lead-lag 超前滞后
leadless chip carrier (LLCC) 无引线芯片载体
leadless package 无引线外壳
lead loss 铅耗
lead module 超前组件
lead pattern 引线图案
lead shield 铅屏
lead wire 引线
leaf 箔;薄片,薄板;弹簧片
leaf electrometer 箔验电计
leaf electroscope 箔验电器
leak 漏泄电阻;漏;漏电
leakage 泄漏,漏液,漏电
leakage check 密封检查
leakage coefficient 漏泄系数
leakage current 泄漏电流,漏泄电流
leakage detector 检漏仪
leakage discharge 泄漏放电
leakage field 漏泄场
leakage flux 漏磁通
leakage inductance 漏感
leakage loss 漏汽损失,漏泄损失
leakage operating current 动作泄漏电流
leakage rate 漏电率
leakage relay 漏电继电器
leakance 漏泄电导

leak detector 检漏仪,漏泄检验器
leak pressure 泄漏压力
leakproof 不漏电的,耐漏液的,防漏的
leaky light guide 漏泄光波导
Leblanc connection 勒布朗克联结
Leblanc exciter 勒布朗克励磁机
Lecher line 勒谢尔线
Lecher wire 勒谢尔线
Leclanché cell 勒克朗谢电池
LED¹ (large electronic display) 大电子显示器
LED² (light-emitting diode) 发光二极管
LEED (low-energy electron diffraction) 低能电子衍射
left-handed rotation 左旋,左向旋转
left-handed screw 左转螺旋
left-handed twist 左扭转
left-hand rule (LHR) 左手定则
leg 支线,引线;变压器铁芯柱
legal ampere 法定安培
legal ohm 法定欧姆
legal volt (LV) 法定伏特
Lenard rays 勒纳德射线
length (l) 长度
length coefficient 长度系数
length of delay 延迟值
length of lay 绞(合节)距
length of span 跨度距离
lengthwise flatness 纵向平直度
Lenz's law 楞次定律
Leonard control 伦纳德控制
leptokurtosis 尖峰态
lepton 轻子
LET (linear energy transfer) 线性能量转移
lethargy 勒
levecon (level control) (信号)电平控制
level adjustment 电平调节
level control (levecon) (信号)电平控制

level diagram 电平图
leveller 水平仪,校平器
level meter (LM) 电平表,电平计
level-shifting amplifier (LSA) 电平漂移放大器
level-shifting diode 电平漂移二极管
level trigger (LT) 电平触发器
lever 杠杆
lever drive 杠杆式传动
lever lock 杠杆锁
lever switch 杠杆开关
Leyden jar 莱顿瓶
LF (low frequency) 低频
LFB (local feedback) 局部反馈,本机反馈
LFC (low-frequency choke) 低频扼流圈
LFF (low-frequency filter) 低频滤波器
LFT¹ (laser flash tube) 激光闪光管
LFT² (linear flash tube) 线性闪光管
LG (loop gain) 环路增益
L grader L 分级器
LHR (left-hand rule) 左手定则
Liapunov function 李雅普诺夫函数
Liapunov's method 李雅普诺夫法
Liapunov's theorem 李雅普诺夫定理
library 库
library searching 谱库检索
LIC (laser image converter) 激光变像器
lidar (激)光雷达
lidar echo (激)光雷达回波
lidar impulse (激)光雷达脉冲
lidar tracking (激)光雷达跟踪
LIF (laser interference filter) 激光干扰滤波器
lifter 升降机,起重机
light-activated SCR (LASCR) 光启可控硅整流器
light-activated silicon switch (LASS) 光启硅开关
light-activated switch (LAS) 光启开关
light-activated thyristor 光控晶闸管
light amplifier 光放大器
light conductor 光导管
light current 弱电流;光电流
light deflection 光偏转
light-dependent resistor (LDR) 光敏电阻
light-emitting diode (LED) 发光二极管
light frequency 光频,光频率
light guide 光导
light guide damping 光导衰减
light guide transmission 光导传输
lighting cable 照明电缆
lighting circuit 照明电路
lighting effectiveness factor 照明有效性因数
lighting feeder 照明馈路
lighting fitting 灯具
lighting load 照明负荷,照明负载
lighting network 照明网络
lighting switch 照明开关
lighting voltage 照明电压
light meter 照度计
light modulator 光调制器
lightning 闪电
lightning arrester 避雷器
lightning conductor 避雷针,避雷线
lightning current 雷电流
lightning current steepness 雷电流陡度
lightning impulse 雷电冲击
lightning overvoltage 雷电过电压
lightning path 雷电路径
lightning rod 避雷针
lightning stroke 雷击
lightning surge 雷电浪涌
light pipe (导)光管

light-positive 正光电导的
light pressure 光压
light quantum 光量子
light relay 光(控)继电器
light resistance 光电阻;耐光性
light-sensitive resistor (LSR) 光敏电阻
light-sensitive tube (LST) 光敏(电子)管
limb 分度弧
lime 石灰,氧化钙
limit capacity 极限容量
limit curve 极限曲线
limited angle torque motor 有限转角力矩电机
limited current circuit 限流电路
limiter 限幅器;限制器
limiter circuit 限幅器电路
limiting breaking capacity 极限分断容量
limiting error 极限误差
limiting resistance 限流电阻
limiting resolution 极限分辨率
limiting value 限值
limit load 极限载荷
limit of interference 干扰限值
limit value 极限值
limpidity 清澈度,透明度
linear accelerator 直线加速器
linear circuit 线性电路
linear energy transfer (LET) 线性能量转移
linear expansion 线性膨胀
linear flash tube (LFT) 线性闪光管
linear integrated circuit 线性集成电路
linearity 线性,直线性;线性度
linearity control 线性控制
linearity error 线性误差
linearizing circuit 线性化电路
linearizing resistance 线性化电阻
linear modulator 线性调制器
linear operator 线性算子,线性算符
linear resistance 线性电阻
linear system simulation 线性系统仿真
linear technology 线性(电路)技术
linear variable transformer (LVT) 线性可变变压器
line busy (LB) 占线
line concentrator (LC) 集线器,线路集中器
line driving amplifier (LDA) 线路激励放大器
line scanning 行扫描
line transmitter (LT) 线路发报机
line terminal 线路端子
line-type modulation (LTM) 线式调制
line voltage monitor (LVM) 线电压监控器
linkage 匝链;链系,联动装置
link circuit (LC) 链式电路
link fuse 熔线片,链熔线
link insulator 拉杆绝缘子
linkwork 链系,联动装置
liquid damper 液体阻尼器
liquid insulated bushing 液体绝缘套管
liquid insulation 液体绝缘
liquid resistor 液体电阻器
LIT (low-impedance transmission) 低阻抗传输
litre (l) 升
live circuit 带电电路
live line 带电线路
live load (LL) 可变负载;工作负载
live part 带电部分;有电部件
LL^1 (live load) 可变负载;工作负载
LL^2 (low-low) 低低,极低
LLCC (leadless chip carrier) 无引线芯片载体
LLR (load limiting resistor) 限制负载电阻(器)
LM (level meter) 电平表,电平计
load adjustment 负载调节
load angle 功角

load capacity 负载能力
load commutation 负载换相
loaded cable (LC) 加感电缆
loaded resonator 加载共振器
loader 装载机;装入程序
loader and unloader 装卸机,装卸桥
load factor 负载系数
loading 加载;装填,上料
loading coil 加感线圈
loading factor 负荷系数
loading resistor 负载电阻(器)
load limiting resistor (LLR) 限制负载电阻(器)
load line 负载线
load pulse 负载脉冲
lobe 波瓣;凸角
lobe frequency 波瓣频率
lobing 天线射束控制,天线扫掠
local battery (LB) 本地电池
local circuit 局部电路
local distribution 局部分布
local feedback (LFB) 局部反馈,本机反馈
local frequency 本振频率
localizer 定位器;航向信标
local loop 本地环路
local oscillator 本机振荡器
locating ring 定位环
location diagram 位置简图
locator 定位装置,定位器
lock chamber 闸室
locked push button 定位式按钮
locked-rotor current 堵转电流
locked-rotor test 堵转试验
lock-in 锁定
locking 锁定
locking circuit 锁闭电路
locking phase 同步相
lock-on 锁位,跟踪锁定
lockout 封锁;锁定
lockout operation 保持操作
lockout relay 保持继电器
lockup 锁
lock-up relay 自保持继电器;闪锁继电器

locus 轨迹
LOF (lowest observed frequency) 最低测得频率
log conversion voltmeter (LCVM) 对数变换伏特计
logger (电子自动)记录器
logging 记录,登记
LOM (laser optical modulator) 激光光学调制器
lone electron 单电子,孤立电子
longitudinal load 纵向负载
longitudinal oscillation 纵向振荡
long-pitch winding 长距绕组
loop 循环;半波;回路,环路
loop current 回路电流
loop gain (LG) 环路增益
loop inductor 单匝感应器
looping-off 解环
loop motor 环流电动机
low-energy electron diffraction (LEED) 低能电子衍射
lowest observed frequency (LOF) 最低测得频率
low frequency (LF) 低频
low-frequency choke (LFC) 低频扼流圈
low-frequency filter (LFF) 低频滤波器
low-impedance transmission (LIT) 低阻抗传输
low-level logic (circuit) 低电平逻辑电路
low-level transmission 低功率传输
low-loss fibre 低损耗光纤
low-low (LL) 低低,极低
low pressure (LP) 低压
low-voltage fuse 低压熔断器
LP (low pressure) 低压
LSA (level-shifting amplifier) 电平漂移放大器
LSI (large-scale integration) 大规模集成电路
LSR (light-sensitive resistor) 光敏电阻

LST (light-sensitive tube) 光敏(电子)管
LT¹ (level trigger) 电平触发器
LT² (line transmitter) 线路发报机
LTM (line-type modulation) 线式调制
lubricant 润滑剂,润滑油
lubricator 润滑器
lug 接线片;凸缘
lumen 流(明)
luminaire dirt depreciation 灯具污垢减光
lumped circuit 集中参数电路
lux (lx) 勒(克斯)
luxmeter 照度计,勒克斯计
lux second 勒秒
LV (legal volt) 法定伏特
LVM (line voltage monitor) 线电压监控器
LVT (linear variable transformer) 线性可变变压器
lx (lux) 勒(克斯)
Lymar 光子铅板

m (metre) 米
ma (milliampere) 毫安
Mach angle 马赫角
Mach cone 马赫锥
Mach effect 马赫效应
machinability 可切削性,机械加工性
machine 机器
machine tool 机床
Mach number 马赫数
Maclaurin expansion 麦克劳林展开
macroblock 宏块
macrocell 宏单元
macrocircuit 宏电路
macrofarad 兆法拉
macro-instruction 宏指令
macromolecule 高分子,大分子
macroscopic constant 宏观常数
macroscopic stress 宏观应力
madistor 磁控等离子体开关;晶体磁控管
MAF (maximum amplitude filter) 最大振幅滤波器
magner 无功功率
magnet 磁体
magnetic chuck 吸盘
magnetic flux 磁通(量)
magnetic friction clutch 磁摩擦离合器
magnetic particle 磁粉
magnetic pole 磁极
magnetic remanence 顽磁
magnetic rigidity 磁刚度
magnetics 磁学
magnetic saturation 磁饱和
magnetic shunt compensation 磁分路补偿
magnetism 磁性;磁力
magnetizability 可磁化性
magnetizer 磁化器,激磁装置
magneto 永磁发电机,磁石发电机
magneto-diode 磁敏二极管
magnetogram 磁图
magnetograph 磁强记录仪
magnetometer 磁强计
magnetomotive force (MMF) 磁动势,磁通势
magneton 磁子
magneto-optics 磁光学
magnetoresistance 磁(致电)阻
magnetoresistor 磁(致电)阻器,磁敏电阻(器)
magnetoscope 验磁器
magnetostriction 磁致伸缩
magnetrol 磁放大器
magnetron 磁控(电子)管
magnification 放大;放大率
magnifier 放大镜
magnistor 磁变管,(电)磁开关
magnitude 量,量级,数量
magnon 磁波子
main arc 主电弧
main circuit 主电路,主回路
main insulation 主绝缘
main lead 母线
main line 干线
mains frequency 电源频率
mains supply 电网供电
mains switch 电源开关
mains voltage 电源电压
maintainability 可维护性,(可)维修度,(可)维修性
maintenance 维护,维修

maintenance factor (MF) 维持因数, 维护系数
main transformer 主变压器
main winding 主绕组
major insulation 主绝缘
majority carrier 多数载流子
majority gate 多数决定门
major lobe 主波瓣
major loop 大半波; 主回路
make-and-break 通断的
make-and-break contact 闭开触点
make-and-break device 断续装置
make contact 动合触头, 动合触点
making capacity 接通能力
making current 接通电流, 关合电流
making operation 接通操作, 关合操作
making switch 接通开关
mandatory implementation 强制实施
mangle 碾压机
manhole 人孔, 检查井
manifold valve 汇流阀
Mannich base 曼尼希碱
manometer (流体)压强计
marginal coefficient 界限系数
marginal frequency 边际频数
marginal relay 定限继电器
margin capacity 备用容量
margin of safety 安全裕度
marked ratio 标定比
marker 标识器
mark frequency 标记频率
marking circuit 标记电路
marking current 符号电流
Markov chain 马尔可夫链
Markov process 马尔可夫过程
Marx generator 马克斯发生器
maser 脉泽, 微波激射(器)
mask 掩模
mask holder 掩模架
mask hole 掩模窗
mask set 掩模组

mass 质量
mass resistivity 质量电阻率
mass spectrograph 质谱(摄谱)仪
mass spectrometer 质谱仪
mass spectrometry (MS) 质谱法
mass spectroscopy 质谱学
mast 天线杆
master chip 主芯片
master core 主线芯
master drive 主令传动
master frequency 主频率
master mask 母版
master motor 主驱动电动机
master relay 主控继电器
master-slave flip-flop 主从触发器
master-slave manipulator 主从机械手, 随动式机械手
master-slave operating system 主从式操作系统
master slice 母片
master switch 主令开关
MAT (micro-alloy transistor) 微合金晶体管
matched filter 匹配滤波器; 匹配筛选器
matched junction 匹配连接
matched waveguide 匹配波导管
matching attenuator 匹配衰减器
matching circuit 匹配电路
matching transformer 匹配变压器
matrix 矩阵; 基体; 凹模
matrix adder 矩阵加法器
matrix addressing 矩阵寻址
matrix circuit 矩阵(变换)电路
matrix display 矩阵显示
matrix inversion 矩阵求逆, 矩阵反演
matrix theorem 矩阵定理
Matthiessen's standard 马希森(铜丝电阻)标准
maximum amplitude filter (MAF) 最大振幅滤波器
maximum capacity 最大容量

maximum continuous rating (MCR) 最大连续定额,最大持续功率
maximum load 最大负荷,最高负荷
maximum voltage 最高电压
maxwell (Mx) 麦克斯韦
Maxwell bridge 麦克斯韦电桥
Maxwell constant 麦克斯韦常数
Maxwell equations 麦克斯韦方程组
Maxwell model 麦克斯韦模型
Maxwell rule 麦克斯韦定则
maz(o)ut 重油
mbar (megabar) 兆巴
MB (megabyte) 兆字节
MC (microcircuit) 微电路,集成电路
MCR (maximum continuous rating) 最大连续定额,最大持续功率
mean availability 平均可用度
mean current 平均电流
mean deviation 平均偏差
mean frequency 平均频率
mean power 平均功率
measuring bridge 测量电桥
measuring relay 量度继电器
mechanical failure load 机械破坏负荷
mechanical torque rate 机械转矩率
mechanics 力学;机械学
mechanism 机构;机制;机理
mechatronics 机械电子学[技术]
median 中位数;中线
medium frequency (MF) 中频
medium-scale integration (MSI) 中规模集成电路
megabar (mbar) 兆巴
megabit 兆比特
megabyte (MB) 兆字节
megacoulomb 兆库仑
megacycle 兆周
megaelectronvolt (MeV) 兆电子伏(特)
megaerg 兆尔格
megafarad 兆法拉
megagauss 兆高斯
megahertz (MHz) 兆赫
megajoule 兆焦(耳)
megametre 兆米
megavar 兆乏
megavolt (Mv) 兆伏(特)
megavolt-ampere 兆伏安
megawatt (Mw) 兆瓦
megger 兆欧表
megohm 兆欧
megohmmeter 兆欧表
membrane mask 薄膜型掩模
memory capacity 存储容量
memory chip 存储芯片
memory register 存储寄存器
mercury arc 汞弧
mercury arc converter 汞弧变流器
mercury arc valve 汞弧阀
mercury cell 汞干电池
mercury relay 水银继电器
mesa 台面
mesa etching 台面刻蚀
mesa transistor 台面晶体管
mesh 网孔
mesh current 网孔电流
mesh emitter 网状发射极
meson 介子
metadyne 磁场放大机
metal arc welding 金属弧焊
metal film resistor 金属膜电阻器
metal foil capacitor 金属箔电容器
metal insulator 金属绝缘体
metallic bond 金属键
metallic return 金属回线
metallic sheath 金属护套
metallization 金属化
metallurgy 冶金学
metal oxide semiconductor (MOS) 金属氧化物半导体
metamagnetism 变磁性
metastable level 亚稳能级

metastable state 亚稳态
meter 计,表
meter constant 仪表常数
meter key 电表键
metre (m) 米
metrology 计量学
MeV (megaelectronvolt) 兆电子伏(特)
MF[1] (maintenance factor) 维持因数,维护系数
MF[2] (medium frequency) 中频
mg (milligram) 毫克
mho 姆欧
mhometer 姆欧表
MHz (megahertz) 兆赫
MIC[1] (microwave integrated circuit) 微波集成电路
MIC[2] (minimum igniting current) 最小点燃电流
mica capacitor 云母电容器
mica insulation 云母绝缘
micro-alloy transistor (MAT) 微合金晶体管
microammeter 微安表
microampere 微安(培)
microbar 微巴
microcircuit (MC) 微电路,集成电路
microcomponent 微型组件
microcoulomb 微库(仑)
microelement 微型元件;微量元素
micro-gap switch 微隙开关
micromanometer 微压计
micromatrix 微矩阵
micromho 微姆(欧)
micromodule 微模块
microohm 微欧(姆)
micropotentiometer 微型电位计
micropulser 微脉冲发生器
microswitch 微型开关
microvolt 微伏(特)
microvoltmeter 微伏表
microwafer 微型晶片
microwatt 微瓦(特)
microwave 微波

microwave emitter 微波发射器
microwave integrated circuit (MIC) 微波集成电路
microwave-pilot protection 微波纵联保护
mid-line 中间线
Miller code 米勒代码
Miller effect 米勒效应
Miller oscillator 米勒振荡器
Miller time base 米勒时基
milliammeter 毫安表
milliampere (ma) 毫安
milligram (mg) 毫克
millilitre (ml) 毫升
millimetre (mm) 毫米
millimho 毫姆(欧)
millimicron 毫微米,纳米
milliohm 毫欧(姆)
milliohmmeter 毫欧表
millivolt (mv) 毫伏(特)
millivoltmeter 毫伏计
milliwatt (mw) 毫瓦(特)
minimum breaking current 最小分断电流
minimum current interrupting rating 断流下限额定值
minimum igniting current (MIC) 最小点燃电流
minimum value 最小值
mini-switch 小型开关
minor axis 短轴
minority carrier 少数载流子
minority carrier current 少数载流电流
minority carrier lifetime 少数载流子寿命
minor lobe 副瓣
minus tapping 负分接
misoperation 误动作
mistermination 端接错误
mis-trip 误跳闸
misuse failure 误用失效
mixed bed 混(合)床
mixer 合路器;混频器
mixer diode 混频二极管
ml (millilitre) 毫升

mm (millimetre) 毫米
MMF (magnetomotive force) 磁动势,磁通势
modulation 调制
modulation transfer function (MTF) 调制传递函数
modulator 调制器
module 模数;模块
modulus 模数,模量
mol (mole) 摩(尔)
molecule 分子
moment 矩,力矩
momentum 动量
monkey (engine) 锤式打桩机
monoblock 单块;单元机组
monobrid 单片组装法
monobrid circuit 单片混合电路
monocrystal 单晶(体)
monomode fibre 单模光纤
monomotor 单电动机,单发动机
monopulse 单脉冲
monoscope 单像管
monostable relay 单稳态继电器
Monte Carlo method 蒙特卡罗法
MOS (metal oxide semiconductor) 金属氧化物半导体
motherboard 母板
motor meter 电动机式仪表
motor pump 电动泵
MPA (multiphoton absorption) 多光子吸收
MS (mass spectrometry) 质谱法
MSI (medium-scale integration) 中规模集成电路
MTF (modulation transfer function) 调制传递函数
muffler 限弧件;消声器
mule 小型电动机车
muller 辗轮混砂机
multi-amplifier 多级放大器
multiband 多频带
multibeam 多光束,多波束
multibreak switch 多断开关
multicavity filter 多腔滤波器
multi-channel peristaltic pump 多通道蠕动泵

multicircuit winding 多路绕组
multicore cable 多芯电缆
multi-coupler 多路耦合器
multidraw 多点取样
multi-element insulator 多元件绝缘子
multi-emitter transistor 多发射极晶体管
multiframe 复帧
multiframe core 多框铁芯
multi-frequency system 多频系统
multi-function measurement instrument 多功能测量仪表
multigang switch 多联开关
multigrid 多栅极
multigrid tube 多栅管
multihead 多传感头
multi-image 多重图像
multilayer 多层
multilayer cathode 多层阴极
multilayer substrate 多层基片
multilead 多引线的
multilevel encoding 多电平编码
multiloop 多回路的;多匝的;多环的
multiloop control system 多回路控制系统
multimeter 万用表,多用表
multimode fibre 多模光纤
multinode 多节点
multi-outlet assembly 多引线装置
multiparameter regulation 多参数调整
multipacting 次级电子倍增
multiphase 多相
multiphase reclosing 多相重合闸
multiphoton absorption (MPA) 多光子吸收
multiphoton process 多光子过程
multiple access 多址
multiple connection 复接,多重连接
multiple electrode 多极,多电极
multiple frequency 倍频,多重频率
multiple operation 并联运行

multiple twin quad 复对四线组
multiple wave winding 复波绕组
multiplex 多路传输,多路复用
multiplexer 复用器
multiplexer filter 复用器滤波器
multiplex lap winding 复叠绕组
multiplication circuit 乘法电路
multiplier 乘数;乘法器;倍增器
multiplier gain 倍增器增益
multiplier phototube 光电倍增管
multiplier tube 倍增管
multiply charged ion 多电荷离子
multipolar machine 多极电机
multipole fuse 多极熔断器
multiposition 多位置的
multiposition element 多位置元件
multiprocessing 多重处理,多道处理
multiprocessing system 多重处理系统
multi-purpose instrument 多功能仪表
multi-purpose oscilloscope 多用示波器
multi-range instrument 多量程仪表
multi-scale instrument 多标度仪表
multiscaler 通用换算器;多路定标器
multisection filter 多节滤波器
multi-speed motor 多速电动机

multistage 多级
multistage amplifier 多级放大器
multiswitch 复接机键,复接开关
multitester 万用表,多用表
multitube stack 多筒式烟囱
multi-turn 多匝的,多线圈的
multi-turn potentiometer 多线圈电位器
multivalued nonlinearity 多值非线性
multivector 多重矢量
multivibrator 多谐振荡器
multivoltmeter 多量程伏特计,多量程电压表
multiway 多路的,多向的
multiway cable 多分支电缆
multiway switch 多路开关,多向开关
multi-winding transformer 多绕组变压器
multi-zone furnace 多区炉
mutation 变异
muting 噪声消减
Mv (megavolt) 兆伏(特)
mv (millivolt) 毫伏(特)
Mw (megawatt) 兆瓦
mw (milliwatt) 毫瓦(特)
Mx (maxwell) 麦克斯韦
Mylar 聚酯树脂
myriabit 万位
myriametric wave 甚长波

na (nanoampere) 纳安(培),毫微安(培)
naked wire 裸线
nanoammeter 纳安计,毫微安计
nanoampere (na) 纳安(培),毫微安(培)
nanocircuit 纳米级电路
nanocomposite 纳米复合材料
nanofarad 纳法,毫微法
nanohenry 纳亨,毫微亨(利)
nanometre (nm) 纳米,毫微米
nanoprocessor 毫微处理器
nanoprogram 毫微程序
nanosecond (ns) 纳秒,毫微秒
nanotechnology 毫微加工技术
nanovolt 毫微伏(特)
nanowatt 纳瓦,毫微瓦(特)
napier 奈培
national super grid 全国超高压电网
natural attenuation 固有衰减
natural damping 自然阻尼
natural period 固有周期
NEA (negative electron affinity) 负电子亲和势
NEA cathode 负电子亲和势阴极
near accident 未遂事故
near echo 近回波
near-end crosstalk 近端串音
near infrared (NIR) 近红外
near ultraviolet (NUV) 近紫外
needle deviation 指针偏转
needle flame test 针焰试验
needle indicator 指针指示器
needle valve 针形阀
negater 反相器,非门
negation 非

negation gate 非门
negative battery 负电池
negative current 负电流
negative electrode 负极,阴极
negative electron affinity (NEA) 负电子亲和势
negative excitation 反向励磁,负励磁,反向激励
negative-glow lamp 负辉光灯
negative image 阴像,负像
negative ion 负离子,阴离子
negative mask 负掩模
negative meson 负介子
negative sequence 负序
negative voltage feedback 电压负反馈
negativity 负性
negatron 负电子,阴电子
negentropy 负熵
neper (Np) 奈培
Nernst bridge 能斯特电桥
Nernst detector 能斯特探测器
Nernst effect 能斯特效应
Nernst-Einstein relation 能斯特-爱因斯坦关系
Nernst lamp 能斯特灯
nesa 奈塞(透明导电膜)
net charge 净电荷
net generation 净发电量
net loss 净损耗
net output 净输出
network analogue 网络模拟
network control (电力)网络控制
network fault (电力)网络故障
network layout (电力)网络布置
neutral bus 中性母线
neutral conductor 中性导体

neutral current 中线电流
neutral earthing 中性点接地
neutral earthing reactor 中性点接地电抗器
neutral grounding 中性点接地
neutrality 中性
neutralization 中性化;中和
neutralizer 中和剂
neutral line 中性线
neutral phase 中性相位
neutral point 中性点
neutral switch 中性线开关
neutretto 中介子
neutrino 中微子
neutron 中子
neutron density 中子密度
neutron doping 中子掺杂
neutron generator 中子发生器
Newton iteration method 牛顿迭代法
Newton-Raphson method 牛顿-拉弗森法
Newton's law 牛顿定律
NF (noise factor) 噪声系数
nibble 四位字节,四位组
Nichols chart 尼科尔斯图
nickel electrode 镍电极
NiFe 镍铁合金
NiFe accumulator 镍铁蓄电池
NiFe cell 镍铁电池
nippers 钳
nipple 螺纹接头
NIR (near infrared) 近红外
nit 尼特
nitrogen 氮
nm (nanometre) 纳米,毫微米
NMR (nuclear magnetic resonance) 核磁共振
nodal admittance matrix 节点导纳矩阵
nodal circle 节圆
nodal force 节点力
node method 节点法
node voltage 节点电压
noise amplitude 噪声振幅
noise clipper 噪声限制器

noise diode 噪声二极管
noise eliminator 消声器
noise factor (NF) 噪声系数
noise meter 噪声计
noise rating (NR) 噪声等级
noise source 噪声源
noise voltage 噪声电压
no-load apparent power 空载表观功率
no-load current 空载电流
no-load loss 空载损耗,空载损失
no-load operation 空载运行
no-load speed 空载转速
no-load tap-changer 空载分接开关
no-load test 空载试验
nominal band 标称频带
nominal capacity 标称容量,额定容量
nominal current 标称电流,额定电流
nominal ratio 标定比
nominal resistance 标称电阻,额定电阻
nominal value 标称值
nominal voltage 标称电压,额定电压
nomogram 诺模图
nomograph 诺模图
nomotron 开关电子管
non-axiality 不同轴性
non-coherent detection 非相干探测
non-coherent radiation 非相干辐射
non-coincidence 不一致
non-combustible construction 不燃结构
non-conductor 非导体,绝缘体
nonconservation 不守恒
non-contact 无接点的,无接触的
non-contact relay 无触点继电器
non-corrodibility 不腐蚀性
non-corrosibility 不腐蚀性
non-defective 良品,合格品
non-delimiter 非定界符

non-detachable part 不可拆件
non-dimensional coefficient 无因次系数,无量纲系数
non-dimensional parameter 无因次参数,无量纲参数
non-eddying flow 无湍流,无涡流
non-electrolyte 非电解质
non-fusibility 抗熔性
non-homing switch 不复位开关,自锁开关
non-inductive capacitor 无感电容器
non-inductive circuit 无感电路
non-inductive resistance 无感电阻
non-inductive shunt 无感分流器
non-inverting amplifier 同相放大器
non-inverting connection 同相连接
non-inverting input 同相输入
nonius 游标
non-laminar beam 非层流粒子束
non-linear dependence 非线性相关
non-linear factor 非线性系数
non-linearity 非线性
non-linear modulation 非线性调制
non-locking key 非锁定电键,自动还原电键
non-locking relay 非锁定继电器,自动还原继电器
non-loss decomposition 无损分解
non-modulation 非调制
non-ohmic resistor 非欧姆律电阻(器)
non-operating value 不动作值
non-orthogonality 非正交性
non-oscillating discharge 非振荡放电
non-parallelism 不平行性
non-pickup 不吸动值
non-polar dielectrics 非极性电介质
non-polarized light 非偏振光
non-polarized relay 非极化继电器
non-reactive power 无电抗功率,有功功率
non-reactive resistance 无电抗电阻
non-relevant failure 非关联失效
non-rewirable plug 不可重接插头
non-salient pole 隐极
non-synchronous motor 异步电动机
non-uniform field 非均匀场,不均匀场
non-uniform insulation 分级绝缘
no-op instruction 空操作指令
NOR circuit 或非电路
NOR gate 或非门
normal axis 法线轴
normal band 基带
normal derivative 法向导数
normal frequency 正常频率
normal impedance 标准阻抗
normality 常态;规度;正态性
normalization 正态化;归一化,规格化,标准化
normalization factor 归一化因子
normalization point 归一化点
normalized curve 标准化曲线
normalized frequency 归一化频率,标准化频率
normalized resistance 归一化电阻,标准化电阻
Norton's theorem 诺顿定理
Norton transformation 诺顿变换
NOT AND circuit 与非电路
notch 凹槽,凹口
notch frequency 陷波频率
notch power meter 陷波功率计
NOT circuit 非电路
NOT gate 非门
no-voltage 无电压,零电压
no-voltage relay 无电压继电器
no-voltage release 无电压释放
noy 纳
nozzle 喷嘴,喷管
nozzle blade 喷嘴导叶

nozzle coefficient 喷嘴系数
nozzle flowmeter 喷嘴式流量计
Np (neper) 奈培
NR (noise rating) 噪声等级
ns (nanosecond) 纳秒,毫微秒
nuclear charge 核电荷
nuclear graphite 核石墨
nuclear laser 核激光器
nuclear magnetic resonance (NMR) 核磁共振
nuclear power 核电
nucleater 成核剂
nucleon 核子
nucleus 原子核,晶核
null adjustment 零位调整
null galvanometer 零点检流计
null measurement 指零测量
null method 零位法

null phase error 零相位误差
null position voltage 零位电压
null voltage 零位电压
numerator 分子;计数器
numeroscope 数字记录器,示数器
nut 螺母,螺帽
nutating feed 盘旋馈电
nutation 章动
nutation field 盘旋场
nutation sensor 章动敏感器
nut switch 螺帽形开关
NUV (near ultraviolet) 近紫外
Nyquist-Cauchy criterion 奈奎斯特-柯西判据
Nyquist curve 奈奎斯特曲线
Nyquist frequency 奈奎斯特频率
Nyquist's theorem 奈奎斯特定理

object distance 物距
objective function 目标函数
objective (lens) 物镜
objective photometer 客观光度计
object module 目标模块
object program library 目标程序库
oblateness 扁圆形;扁率
oblique anode 倾斜阳极
oblique deposition 倾斜淀积
obliquity 倾斜;斜度
observability 可观测性
observation 观测,观察
observation circuit 监视电路
observation frequency 观测频率
observed reliability 观测可靠度
observer 观察器,观测器
obstacle 阻挡物,障碍物,阻碍
obstacle detection 障碍物探测
obstacle diffraction 障碍衍射
obstacle gain 障碍增益
obstacle indicator 故障指示器
OC (open circuit) 开路,断开电路
OCB¹ (oil circuit-breaker) 油断路开关,油断路器
OCB² (overload circuit-breaker) 过载断路器
OCD (optically coupled device) 光耦合器件
OCR (optical character recognition) 光(学)字符识别
octal digit 八进制数字
octal notation 八进制计数法
octode 八极管
odd function 奇函数
odd harmonic function 奇调和函数

odd symmetry 奇对称
Oe (oersted) 奥斯特
OEIC (optoelectronic IC) 光电子集成电路
off condition 截止状态
off contact 触点断开
off resistance 关态电阻
offset 失调,偏移,偏移量;偏置;补偿
offset beam 偏移射束
offset current 补偿电流;偏移电流,失调电流
offset frequency 偏频,偏移频率
offset voltage 补偿电压;偏移电压,失调电压
off state 断态,断开状态
off-state current 断态电流
off-state voltage 断态电压
off transistor 截止晶体管
ohm 欧(姆)
ohmage 欧姆电阻,欧姆阻抗
ohm ammeter 欧安表
ohmic bridge 欧姆电桥,电阻电桥
ohmic contact 欧姆接触
ohmic drop 电阻电压降
ohmic junction 欧姆接合
ohmic leakage 漏电阻
ohmic loss 电阻损失,欧姆损失
ohmic resistance 欧姆电阻
ohmmeter 欧姆计,电阻表
Ohm's law 欧姆定律
oil brake 油制动器
oil circuit-breaker (OCB) 油断路开关,油断路器
oil condenser 油浸电容器
oil cooler 冷油器

oil damping 油阻尼
oil feeder 受油器
oil-filled pipe-type cable 管式充油电缆
oil hydraulic motor 油压马达
oil insulation 油绝缘
oil-jacked bearing 油顶起轴承
oil switch 油开关
oil transfer pump 输油泵
omegatron 回旋质谱计
omni-bearing indicator 全方位指示器
omnidirectional antenna 全向天线
omnidirectional radiation 全向辐射
omnidirectional reception 全向接收
omnitron 全能加速器
OMR (optical mark reader) 光(学)标记读取器
one-address code 单地址码
one-chip amplifier 单片放大器
one-fluid cell 单液电池
one-loop system 单环路系统
on-load operation 有载运行
on-load factor 负载因数
on-load switch 负荷开关
on-load test 带负荷试验
on-load voltage 有载电压
on-load voltage regulation 有载调压
on-off action 通断作用
on-resistance 导通电阻
on-state 通态,接通状态
on-state current 通态电流
on-state power dissipation 通态耗散功率
on-state voltage 通态电压
opacity 不透明度
opaque plasma 不透明等离子体
opaque region 不透明区
OPD (optical path difference) 光程差
open circuit (OC) 开路,断开电路
open-circuit arc 开路电弧
open-circuit spinning 开路自转
open-circuit test 开路试验
open-circuit voltage 开路电压
opening 断开;缝隙,孔口
open lapping 间隙绕包
open-loop control 开环控制
open-loop system 开环系统
open-phase relay 断相继电器
open-phase running 断相运行,非全相运行
open-type resonator 开启式谐振腔
open-wire circuit 明线线路
operating voltage 工作电压
operation 操作;运转;运算
operational amplifier 运算放大器
operational research 运筹学
operation code 操作码
OPO (optical parametric oscillator) 光参量振荡器
opposed firing 对冲燃烧
opposite electricity 异性电
opposite phase 反相
opposition method 反接法,对消法
optical actuator 光致动器
optical ammeter 光学安培计
optical beam splitter 光束分束器
optical character reader 光(学)字符读取器
optical character recognition (OCR) 光(学)字符识别
optical circulator 光学环行器
optical conductor 光导体
optical coupling 光耦合
optical crystal 光学晶体
optical deflector 光偏转器
optical encoder 光编码器
optical fibre communication 光纤通信
optical fibre concentrator 光纤集中器
optical fibre scattering 光纤散射
optical fibre splice 光纤接头
optical image formation 光学成像

optical injection 光注入
optical integrated circuit 光集成电路
optical lever 光杠杆
optically coupled device (OCD) 光耦合器件
optical mark reader (OMR) 光(学)标记读取器
optical mode 光学模
optical modulator 光调制器
optical multiplexer 光学(多路)复用器
optical parametric oscillator (OPO) 光参量振荡器
optical path difference (OPD) 光程差
optical pickup 光碟(读)头,光学传感器
optical pump 光泵
optical radar 光学雷达
optical relay 光继电器
optical resonance 光学谐振
optical storage 光学存储(器)
optical waveguide 光波导
opticity 旋光性
optics 光学
optimal control 最优控制
optimal criterion 最佳判据
optimal damping 最佳阻尼
optimal efficiency 最佳效率
optimal parameter 最佳参数
optimization (最)优化
optimizer 优化程序;最优控制器
optimum allocation 最佳配置,最优分配
optimum coupling 最佳耦合
optimum output 最佳输出
option 选择,选项,选件
optional equipment 附加设备,任选设备
option switch 选择开关
optiphone 特种信号灯
opto-coupler 光耦合器件
optoelectronic amplifier 光电子放大器
optoelectronic chip 光电子芯片
optoelectronic IC (OEIC) 光电子集成电路
opto-isolator 光隔离器;光绝缘体
optotransistor 光晶体管
optronics 光电子学
OR 或
orbit 轨道;范围
orbital angular momentum 轨道角动量
orbital electron 轨道电子
orbital function 轨道函数
OR circuit 或电路
OR element 或元件
OR function 或函数;或功能
organic ester 有机酯
organic semiconductor 有机半导体
OR gate 或门
oriented crystallization 定向结晶
orifice 孔口
orifice flow 孔口出流
orifice flowmeter 孔板流量计
O-ring O形环
O-ring gasket O形垫圈
O-ring seal O形环密封
OR operator 或算子
orotron 奥罗管
OR switch 或开关
orthoaxis 主轴,正交轴
orthocentre 垂心
orthoclase 正长石
ortho-compounds 邻位化合物
orthodiagraph 正摄像仪
orthogonality 正交性
orthogonal lattice 直角点阵
orthogonal matrix 正交矩阵
orthogonal polarization 正交极化
orthogonal scanning 正交扫描
orthograph 正视图,正投影图
orthography 正投影,正投射法
orthonormality 规格化正交性,标准正交性
orthopole 正交极;垂极
orthoscanner 垂向扫描器
OR tube 或门管
oscillating arc 振荡电弧

oscillating circuit 振荡电路
oscillating coil 振荡线圈
oscillating discharge 振荡放电
oscillating frequency 振荡频率
oscillating function 振荡函数
oscillating klystron 振荡速调管
oscillating period 振荡周期
oscillating voltage 振荡电压
oscillation 振荡,振动;摆动
oscillation damping 振荡阻尼
oscillation effect 振荡效应
oscillation generator 振荡发生器
oscillation tube 振荡管
oscillator 振荡器
oscillight 显像管
oscillion 三极振荡管
oscillistor 半导体振荡器
oscillogram 波形图;示波图
oscillograph 录波器,示波器
oscillography 示波术,示波法
oscillometer 示波计
oscilloprobe 示波器探头
oscilloscope 示波器,录波器
oscilloscope tube 示波管
oscillosynchroscope 同步示波器
O-scope O 型显示器
osmium 锇
osmosis 渗透
osmosis tube 渗透管
osram 锇钨灯丝合金;灯泡钨丝
O-type device O 型器件
O-type tube O 型管
ounce 盎司
outage 运转中断期,停机
outlet line corridor 出线走廊
out-of-balance current 不平衡电流
out-of-frame 帧失调
out-of-line coding 线外编码,越线编码
out of phase 失相
out-of-phase current 异相电流
out of step 失步
out-of-step blocking 振荡闭锁
out-of-step protection 失步保护
out-of-step relay 失步继电器
out-of-step tripping 失步跳闸
out of synchronism 失步
output circuit 输出电路
output device 输出装置,输出设备
output immittance 输出导抗
output make circuit 动合输出电路
overbias 过偏压
overcapacity 过容量
overcharge 过充电
overcompounded motor 过复励电动机
overcompounding 过复励,超复励
over-coupled transformer 过耦合变压器
over-coupling 过耦合
overcurrent 过电流
overcurrent blocking device 过电流闭锁装置
overcurrent discrimination 过电流鉴别
overcurrent diverter 过电流分流器
overcurrent protection 过电流保护
overcurrent protective coordination 过电流保护配合
overcurrent relay 过电流继电器
overcurrent release 过电流脱扣器
overdamping 过阻尼
over-excitation 过激励,过励磁
overhead cable 架空电缆
overhead conductor 架空导线
overhead line 架空线路
overheat 过热
overheating protection 防超温保护,过热保护
overheat relay 过热继电器
overload 过载,超载,过负荷
overload circuit-breaker (OCB) 过载断路器
overload margin 过载容限,过载裕度

overload relay 过载继电器
overload trip 过载跳闸
overload valve 过载阀
overpotential 超电势,过电位
overpressure disconnector 过压力隔离器
overreaching protection 过范围保护
overrun 超过限度;溢流
overvoltage 过电压
overvoltage protection 过电压保护
overvoltage relay 过压继电器
overvoltage suppressor 过电压抑制器
oxidant 氧化剂
oxidation 氧化
oxide 氧化物
oxidizer 氧化剂
oxygenation 氧合作用
oxygen blow 吹氧
ozone 臭氧
ozone generator 臭氧发生器
ozonizer 臭氧发生器

P (poise) 泊(动力黏度单位)
PA (polyamide) 聚酰胺
Pa (pascal) 帕(斯卡)
packed-type solenoid valve 填料函型电磁阀
packet mode 包方式
packet sequencing 包排序
packet switching 包交换
packing 密实度
packing box assembly 填料函组件
packing machine 装填机
P-action 比例作用,P作用
pad 衬垫;焊盘
page addressing 页面寻址
page-at-a-time printer 一次一页式打印机,页式打印机
page printer 页式打印机
pan charger 盘式装料装置
P & ID¹ (piping and instrument diagram) 管道及仪表流程图
P & ID² (process and instrument diagram) 工艺及仪表流程图
panel 面板;节间
pantograph 受电弓
parabola 抛物线
parallax 视差
parallel capacitive compensation 并联电容补偿
parallel circuit 并联电路
parallel computer 并行计算机
parallel connection 并联
paralleling 并联(运行)
parallel inverter 平行逆变器
parallelism 并行性;平行度
parallel operation 并行操作;并联运行

parallel output system 并行输出制
parameter 参数,参量
partial pressure 分压力
partial volume 分体积
particulate 微粒
partition 分区;隔开物
parts list (PL) 元件表,零部件清单
pascal (Pa) 帕(斯卡)
passing contact 滑过触点
passivation 钝化
passive circuit 无源电路
pasted plate 涂膏式板板
path 通路,路径
PCB (printed circuit board) 印刷电路板
p-channel FET p沟(道)场效(应)晶体管
PCM (pulse code modulation) 脉码调制,脉冲编码调制
p contact p型接触
PCV (pressure control valve) 压力控制阀
p diffused region p型扩散区
p doped drain p型掺杂漏极
p doped source p型掺杂源极
PE (power exchange) 电力交易;功率交换
peak factor 峰值因数
peak-load duty 峰值负载工作制
peak-load plant 尖峰负荷电站
peak (value) 峰值
peak voltmeter 峰值电压表
Peltier effect 佩尔捷效应
PEN conductor 保护中性导体
pendant 吊灯

pendant fitting 吊灯(具)
penetration 穿透;贯入件;贯入度
penstock 水闸;压力管道,压力钢管
pentane 戊烷
penultimate stage 次末级
percolator 渗滤咖啡壶
perfect gas 完全气体
perfect vacuum 完全真空
performance band 性能频带
peripheral device 外围设备控制器
peripheral equipment 外围设备
peripheral flow rate 周缘流量
peripheral interface adaptor (PIA) 外围接口适配器
periphery 外围
permanent fault 永久性故障
permanent magnetic lens 永磁透镜
permanent-magnet moving coil galvanometer 永磁动圈式[磁电系]检流计
permanent-magnet moving coil instrument 永磁动圈式[磁电系]仪表
permeability 渗透率;磁导率
permeability tube 渗透管
permeameter 磁导计
permillage 千分率
permissible cumulative discharge of standard cell 标准电池允许累计放电量
permissible discharge of standard cell 标准电池允许放电量
peroxide 过氧化物
petrochemical 石油化学品
petroleum 石油制品
petroleum extraction 石油精炼
petroleum oil detection buoy system 石油探测浮标系统
phase 相,相位
phase difference 相(位)差
phase displacement 相(位)移
phase mark 相位标记
phase winding 相绕组

pH-indicator pH 指示器
photodetection 光(电)探测
photoelectric cell 光电池
photoelectricity 光电
photoelectron 光电子
photolysis 光解作用
photometer 光度计,测光仪
photon 光子
photoneutron 光中子
phototube 光电管
PIA (peripheral interface adaptor) 外围接口适配器
pickup 传感器;拾音器,检波器
piecewise test 分段电压试验
piezoelectricity 压电
piezometer 测压管,测压计
pilot arc 引导电弧
pilot exciter 副励磁机
pilot-plant stage 试点电站阶段
pilot protection 纵联保护
pin 插销;插针
pin ball 脚球
pinion 小齿轮
pipe cooling 水管冷却
pipeline flushing 冲管
pipeline loss 沿程损失
pipe-type cable 管式电缆
pipework 管路,管道系统
piping 管网,管道,管系;管涌
piping and instrument diagram (P & ID) 管道及仪表流程图
piston 活塞
pitch factor 节距因数
pitch diameter 螺纹中径,节圆直径
pit liner 凹槽衬垫
Pitot tube 皮托管,空速管
pitting 点蚀斑
pivot support frame 枢轴支架
PL (parts list) 元件表,零部件清单
placement 布局
placement of boiler drum 汽包就位
plan 平面图
plane 平面

planetary electron 轨道电子
planimeter 求积仪,面积仪
plank 厚板
plasma 等离子体
plasma arc 等离子弧
plasma spraying 等离子喷涂
plasticizer 增塑剂
plate 板极
plate fuel assembly 板状燃料组件
plate heat exchanger 板式换热器
platen 电极台板
platinum spiral filament 螺旋铂灯丝
plug-in circuit-breaker 插入式断路器
pneumatic actuator 气动执行机构
pneumatic angular displacement actuator 角行程气动执行机构
pneumatic control 气动控制
pneumatic limit operator 气动极限操作器
pneumatic piston gauge 气动活塞式压力计
pneumatic pump 气动泵
pneumatic rabbit 气动跑兔
pneumatic surge chamber 气压式调压室
pneumatic system 气压系统,气动系统
pneumatic thickness meter 气动厚度计
point contact 点接触
point drift 点漂
pointer 指针
pointer adjustment 指针调整
pointer galvanometer 指针式检流计
pointer instrument 指针式仪表
pointer length 指针长度
point filament 点头灯丝
point location 点定位
point-to-point connection 点对点连接
point-to-point control 点对点控制,点位控制
point-to-point control system 点到点控制系统
point-to-point resolution 点分辨力
point-to-point transmission 点对点传输
poise (P) 泊(动力黏度单位)
polarimeter 旋光仪
polarity 极性
polarization charge 极化电荷
pole face (磁)极面
pole pitch 极距
pole shoe 极靴
pollutant 污染物
polyamide (PA) 聚酰胺
polymer 聚合物
pondage 水库贮水量
porcelain bushing 瓷套管
porcelain cleat 瓷夹板
porcelain insulator 瓷绝缘子
pore 气孔,小孔
porosity 孔隙度
position encoder 位置编码器
positioner 定位器
position error 位置误差
position error coefficient 位置误差系数
position feedback 位置反馈
position indicating switch 位置指示开关
position measuring instrument 位置测量仪
position sensor 位置传感器
position transducer 位置传感器
positive displacement flowmeter 容积式流量计
positive feedback 正反馈
positive-negative action 正负作用
positive-negative three-step action 正负三位作用
positive pressure 正压
positive shock response spectrum 正冲击响应谱
positive strain 正应变

positive system 正系统
positive temperature coefficient thermistor 正温度系数热敏电阻
positive terminal 正极柱
potassium 钾
potential difference 电位差
potential energy 势能,位能
potentiometer 电位器,电位差计
potentiometer compass 电位器罗盘
potentiometer (type) pressure transducer 电位器式压力传感器
potentiometric analyzer 电位式分析器
potentiometric displacement transducer 电位器式位移传感器
potentiometric sensor 电位器式传感器
potentiometric transducer 电位器式传感器
potentiometry 电位(滴定)法
power cable 电力电缆
power-compensation differential scanning calorimetry 功率补偿型差示扫描量热法
power exchange (PE) 电力交易;功率交换
power factor meter 功率因数表
power frequency 工频
powerhouse 发电站;电站厂房
power loss 功率损耗
power source 电源
power source welding voltage 电源焊接电压
power spectral density 功率谱密度
power station 电厂
power stroke 做功冲程
power supply device 电源装置
power supply frequency 电源频率
power supply voltage 电源电压
power system automation 电力系统自动化
Prandtl number 普朗特数
preamplifier 前置放大器
precipitation hardened alloy 脱溶硬化合金
predicted reliability 预计可靠度
preferred orientation 择优取向,优先定向
premium fuel 优质燃料
pressure control valve (PCV) 压力控制阀
pressured fluidized-bed combustion 增压流化床燃烧
pressure loss 压力损失
pressure reducing valve 减压阀
pressure turbine 压力式涡轮
pressurized enclosure 正压外壳
pressurized water reactor (PWR) 压水(反应)堆
pre-start checking 启动前检查
prestress 预应力
preventive maintenance 预防性检修,预防性维护
PRF (pulse repetition frequency) 脉冲重复频率
primary air 一次风
primary cell 原电池,一次电池
primary winding 初级绕组,一次绕组
prime mover 原动机
priming 汽水共腾;引火;底漆;雷管
principal current 主电流
principal tapping 主分接
printed circuit board (PCB) 印制电路板
probe coil 探头线圈
probe coil clearance 探头线圈间隙
probe ion 探针离子
probe method 探针法
probe microphone 探管传声管
probe rotational scan 探头摆动扫查
probe-to-flaw distance 探头-缺陷距离

process and instrument diagram (P & ID) 工艺及仪表流程图
process control 过程控制
process gas chromatograph 流程气相色谱仪
processibility test 工艺性能试验
process I/O 过程输入输出
process liability 过程责任
process mass spectrometer 流程质谱计
process measurement 过程测量
process model 过程模型
processor 处理器,处理机,处理单元
process-oriented sequential control 过程定序顺序控制
process-oriented simulation 面向过程仿真
prod 支杆触头
prod method 支杆法
propelling movement 推送运行
proportional control 比例控制
proportional controller 比例控制器
protected creepage distance 保护爬电距离
protected heating element 防护式加热元件
protection factor 保护因数
protective current transformer 保护电流互感器
protective margin 保护裕度
proton 质子
prototype 原型
protruding pole 凸极
p-type dopant p型掺杂剂
p-type semiconductor p型半导体
pull-in torque 牵入转距
pull-through winding 拉入绕组
pulsation 脉动
pulse code modulation (PCM) 脉码调制,脉冲编码调制
pulse control 脉冲控制
pulsed Fourier transform NMR 脉冲傅立叶变换核磁共振法
pulsed Fourier transform NMR spectrometer 脉冲傅里叶变换核磁共振波谱仪
pulse duration 脉冲持续时间,脉冲宽度
pulse duration modulation 脉宽调制,脉冲宽度调制
pulse echo method 脉冲回波法
pulse flip angle 脉冲回转角
pulse frequency modulation control system 脉冲调频控制系统
pulse input 脉冲输入
pulse pile-up 脉冲堆积
pulse polarograph 脉冲极谱仪
pulse position modulation 脉冲位置调制
pulse repetition frequency (PRF) 脉冲重复频率
pulse sensor 脉搏传感器
pulse transducer 脉搏传感器
pulse-type airborne electromagnetic instrument 脉冲式航空电磁仪
pulse width 脉冲宽度
pulse-width modulated inverter 脉宽调制逆变器
pump 泵
pumped heater 带泵加热器
pumped storage station 抽水蓄能电站
puncture voltage 击穿电压
pusher operation 补机推送
push-through winding 插入绕组
PWR (pressurized water reactor) 压水(反应)堆
pyrite 黄铁矿
pyrites 硫化铁矿类

QBS (quality-based selection) 根据质量选择
QCB (queue control block) 队列控制块
QDTA (quantitative differential thermal analysis) 定量差热分析
QI (quality index) 品质指标,质量指标
QMS (quality management system) 品质管理体系,质量管理体系
quadding 星绞
quadrant 象限;象限仪
quadrant electrometer 象限静电计
quadrature-axis component 交轴分量
quadrature-axis winding 交轴绕组
quadrupole field 四极场
quadrupole ion trap 四极离子阱
quadrupole magnet 四极磁铁
quadrupole mass filter 四极滤质器
quadrupole mass spectrometer 四极质谱计
quadrupole probe 四极探头
quadrupole residual gas analyzer 四极残余气体分析器
quadrupole rod 四极杆
quadruple valve 四重阀
qualification standard 鉴定标准
qualification test 鉴定试验
qualified facilities 限定设备
qualitative analysis 定性分析
qualitative physical model 定性物理模型
quality-based selection (QBS) 根据质量选择
quality index (QI) 品质指标,质量指标
quality loop 品质环
quality management system (QMS) 品质管理体系,质量管理体系
quantification 定量化
quantitative differential thermal analysis (QDTA) 定量差热分析
quantitative differential thermal analyzer 定量差热分析仪
quantity 数量
quantity of heat 热量
quantity of illumination 光照量
quantity of radiant energy 辐射能量
quantity of radiation 辐射量
quantization 量化
quantum 量子
quarry 采石厂
quartz 石英
quartz thermometer 石英温度计
quasi-equilibrium theory 准平衡理论
quasilinear characteristics 准线性特性
quasi-molecular ion 准分子离子
quasi-peak 准峰值
quasi-peak detector 准峰值检波器
quasi-periodic vibration 准周期振动
quasi-rigid rotor 准刚性转子

quasi-sinusoidal quantity 准正弦量
quasi-static unbalance 准静不平衡
quasi-synchronization 准同步
quench circuit 熄灭电路,灭弧电路
quench cooling phase 骤冷阶段
quench crack 淬火裂纹
quench drum 骤冷箱
quenched spark gap 猝熄火花隙
quencher 猝灭剂;冷渣设备
quench frequency 猝熄频率
quench front 骤熄前沿
quenching 淬火;熄灭,猝熄
quenching circuit 猝熄电路,消火花电路
quenching compound 冷却介质,冷却液
quenching effect 猝灭效应
quenching medium 冷却介质
quenching moment 灭弧时刻,灭弧点
quenching of excitation 激励猝灭
quenching of photoconductivity 光电导性猝熄
quenching pulse 消隐脉冲
quenching resistance 灭弧电阻
quenching system 冷风系统
quenching tank 淬火冷却槽
quenching time 灭弧时间
quenching transformer 猝熄变压器
quenching voltage 熄灭电压
quenching water 冷却水

queue algorithm 排队控制算法
queue control block (QCB) 队列控制块
queued access method 排队存取法
queued sequential access method 排队顺序存取法
queueing theory 排队论
quibinary code 五二码
quick action 速动
quick-action contact 快动作触头
quick-break fuse 快断熔断器
quick-break switch 快断开关
quick dump rinser 高效率冲洗器
quick hardening 快速硬化
quick-operating circuit-breaker 速动断路器
quick-operating relay 速动继电器
quick operation 速动
quick-releasing relay 速释继电器
quick response excitation 快速励磁
quick set-up 快速调节
quiescent current 静态电流
quiescent point 静点
quinhydrone electrode 醌氢醌电极
quinhydrone half cell 醌氢醌半电池
quota 定额,限额
quotient 商
quotient meter 商值表
quotient relay 商继电器

radial field cable 分相屏蔽电缆
radial thickness of ice 复冰厚度
radian 弧度
radiance temperature 亮度温度
radiance thermometry 亮度测温法
radiant element 辐射元件
radiant energy 辐射能
radiant exitance 辐射出射度
radiant flux 辐射通量
radiant intensity 辐射强度
radiant power 辐射功率
radiation 辐射
radiation balance meter 辐射平衡表
radiation detector 辐射探测器
radiation dose sensor 射线剂量传感器
radiation dose transducer 射线剂量传感器
radiation efficiency 辐射效率
radiation flux 辐射通量
radiation heat flowmeter 辐射热流计
radiation intensity 辐射强度
radiation monitor 辐射监测仪
radiation pyrometer 辐射高温计
radiation sensor 辐射传感器
radiation temperature 辐射温度
radiation temperature sensor 辐射温度传感器
radiation temperature transducer 辐射温度传感器
radiation test 辐射试验
radiation thermometer 辐射温度计
radiation thermometry 辐射测温法
radiation transducer 辐射传感器
radioactive contamination 放射性污染
radioactive survey 放射性测量
radio chromatography 放射色谱法
radio direction finder 无线电测向仪
radio frequency cable 射频电缆
radio frequency electrode 射频电极
radio frequency sensor 射频敏感器
radiometer 辐射仪,辐射表
radiosonde 无线电探空仪
radio theodolite 无线电经纬仪
radio tube （无线电）电子管
radius 半径
radon 氡
Rankine degree 朗肯温度
rare earth element 稀土元素
rated burden 额定负载
rated capacity 额定容量
rated continuous thermal current 额定连续热电流
rated dynamic current 额定动态电流
rated flow coefficient 额定流量系数
rated instrument limit primary current 仪表额定极限一次电流
rated load 额定载荷,额定负载
rated operating condition 额定工况
rated operating range 额定工作范围
rated operating specification 额定工作规程

rated output 额定出力,额定输出
rated primary current 额定一次电流
rated primary voltage 额定一次电压
rated range 额定范围
rated range of use 额定使用范围
rated secondary current 额定二次电流
rated secondary voltage 额定二次电压
rated short-time thermal current 额定短时热电流
rated sine excitation force 额定正弦激振力
rated static longitudinal load 额定静态纵向负荷
rated static transverse load 额定静态横向负荷
rated transformation ratio 额定变比
rate-of-rise suppressor 上升率抑制器
rating 额定值,定额;额定功率;等级
RBS (Rutherford backscattering spectroscopy) 卢瑟福背散射谱法
reactance coil 电抗线圈
reactance relay 电抗继电器
reactance voltage 电抗电压
reaction 反应;反作用力
reactive power 无功功率
reaction turbine 反动式汽轮机,反力涡轮
reactor 电抗器;反应堆
realignment 重组
Réaumur scale 列氏度
reboiler 再沸器,重沸器
rechargeable battery 可充电电池
reciprocal 倒数
reciprocating compressor 往复式压缩机
rectangular impulse current 方波冲击电流
rectification 整流

rectifier 整流器
rectifier transformer 整流变压器
recuperator 回流换热器
recycle chromatography 循环色谱法
redox electrode assembly 氧化还原复合电极
redox potential meter 氧化还原电位测定仪
reduced model 降阶模型
reduced order observer 降阶观测器
reduced power tapping 降容量分接
reduced viscosity 还原黏度
reducing plane table equipment 归算平板仪
reduction of area 面缩率,断面收缩率
reduction of length 缩短率
redundancy computer system 冗余计算机系统
redundancy unit 冗余单元,冗余设备
redundant information 冗余信息
reel 卷轴
reentrant 凹角
reentry control 再入控制
reference block 对比试块,参考试块
reference chart 参比记录纸
reference coil 参考线圈,基准线圈
reference column 参比柱
reference condition 参比条件
reference electrode 参比电极
reference junction 参比接点
reference specimen 参考样本
reflection coefficient 反射系数
refractive index 折射率,折射指数
refrigeration 制冷
regenerative cycle 回热循环
regenerator 再生器;蓄热器
reinforced concrete 钢筋混凝土
relative permittivity 相对电容率

relay 继电器;伺服电(动)机
relaying protection 继电保护
reliability 可靠性,可靠度
relief valve (安全)泄放阀,空放阀,降压阀
reluctance 磁阻
re-make time 重接通时间
removable element 可拆元件
renewable energy (可)再生能源
renewable fuse-link 可更换熔断体
representative overvoltage 代表性过电压
repulsion induction motor 推斥感应电动机
reserve cell 储备电池
reservoir 储热层
residual buoyancy 剩余浮力
residual carrier 残余载波
residual current 剩余电流,残余电流
residual current transformer 剩余电流互感器
residual deflection 残余偏转
residual electromotive force of potentiometer 电位差计残余电动势
residual error 残差
residual error rate 残错率
residual fuel oil 残渣油
residual inductance of resistance box 电阻箱残余电感
residual magnetic field detector 剩磁检查仪
residual magnetism 剩磁
residual method 剩磁法
residual resistance of resistance box 电阻箱残余电阻
residual standard deviation 残余标准偏差
residual unbalance 剩余不平衡
residual variance 残余方差
residual voltage 剩余电压
residual voltage at zero 零点残余电压
residual voltage transformer 剩余电压互感器
residual voltage winding 剩余电压绕组
resin 树脂,松脂
resin-bonded graphite brush 树脂黏合石墨电刷
resin-impregnated paper bushing 胶浸纸套管
resistance 电阻
resistance balance 电阻平衡
resistance box 电阻箱
resistance box with plug 插头式电阻箱
resistance box with rotary switch 旋转开关式电阻箱
resistance box with terminals 端接式电阻箱
resistance hygrometer 电阻湿度计
resistance inclinometer 电阻式测斜仪
resistance psychrometer 电阻干湿球湿度计
resistivity 电阻率
resistor 电阻(器)
resolution 分辨率,分辨力
resonance 共振,谐振;共鸣
restrictor 限流阀,节流器
return air duct 回风管道
return cable 回流电缆
return circuit 回流电路
return outlet 回风风口
reverse bias 反向偏压
reverse blocking diode thyristor 反向阻断二极晶闸管
reversing valve 换向阀
reverting value 回复值
revolution 旋转
rewirable plug 可重接插头
Reynolds number 雷诺数
rigid coupling 刚性联轴器
rigidity 刚性,刚度
rigid rotor 刚性转子
ripple 纹波
ripple voltage 纹波电压
rivet 铆钉;铆接

robust control 鲁棒控制
robust controller 鲁棒控制器
robustness 鲁棒性
rock densitometer 岩石密度计
rocketsonde 火箭探空仪
rock press-in sclerometer 岩石压入硬度计
roller 轧辊
root mean square 均方根
rotary compressor 回转压缩机
rotary retort furnace 转罐式炉
rotating disc electrode 转盘电极
rotor 转子；风轮
rotor winding machine 转子绕嵌机
rotor yoke 转子磁轭
runner 转轮
runner cone 泄水锥
runner hub 转轮体
running speed 运行转速
rupture pressure 破坏压力
rust-preventive packaging 防锈包装
rustproof packaging 防锈包装
Rutherford backscattering spectroscopy (RBS) 卢瑟福背散射谱法

safe logic assembly 安全逻辑装置
safe shutdown 安全停堆
safety circuit 安全电路
safety interlock system 安全联锁系统
safety rod 安全棒
safety valve (SV) 安全阀
safety washing distance 安全冲洗距离
SAFP (shop-assembled fabricated piece) 车间组装制作件
sag 垂度,弛度
salient pole 凸极
salinization 盐化作用
sampling action 采样作用
sampling control 采样控制
sampling controller 采样控制器
sampling control system 采样控制系统
sampling cross-section 采样横截面
sampling element 采样元件
sampling error 采样误差
sampling frequency 采样频率
sampling holder 采样保持器
sampling interval 采样间隔
sampling period 采样周期
sampling pulse 采样脉冲
sampling radiometer 取样辐射仪
sampling rate 采样速率
sampling system 采样系统
sampling test 抽样试验
sampling time 采样时间,取样时间
sampling value 采样值
saturated calomel electrode 饱和甘汞电极
saturated standard cell 饱和式标准电池
saturation 饱和
saturation characteristics 饱和特性
saturation deficit 饱和差
saturation vapour pressure 饱和水汽压
scale analysis 水垢分析
scale plate 分度盘
scale prevention treatment 防垢处理
scanner 扫描器
scanning 扫描
scanning electron microscope (SEM) 扫描电子显微镜
scanning ion microprobe 扫描离子微区探针
scanning microwave multiband radiometer (SMMR) 多波段扫描微波辐射仪
scanning radiation thermometer 扫描辐射温度计
scanning transmission electron microscope (STEM) 扫描透射电子显微镜
scanning tunnelling microscope (STM) 扫描隧道显微镜
scan rate 扫描速率
scattered radiation 散射辐射
scatterer 散射体
scattering 散射
scattering ion energy 散射离子能量
scattering ion energy ratio 散射离子能量比值

scattering turbidimeter 散射光浊度计
scatterometer 散射计
scavenge 清除;排除废气
scavenger 清除剂
scheduling 调度
scintillation detector 闪烁探测器
scintillation emanometer 闪烁室式射气仪
scintillator 闪烁体
sclerometer 硬度计
Scleroscope 回跳硬度计,肖氏硬度计
SCR (silicon controlled rectifier) 可控硅整流器
scram 紧急停堆
scrambler 扰码器
scraper 铲运机;清管器;刮刀,刮板
scrap (metal) 废金属
screen 格网;屏蔽
screening effectiveness 屏蔽效能
screw base 螺口灯头
screw cap 螺口灯头
screw compressor 螺杆式压缩机
screw conveyor furnace 螺旋输送式炉
scriber 划针;刻图仪
scroll 涡带
scrubber 洗涤塔,气溅器
scum 浮渣,浮垢
SDD (system design description) 系统设计说明
sealing 密封;封口;止水;止漏环
sealing oil 密封油
secondary cell 蓄电池
secondary coating 二次被覆层
secondary combustion 再燃烧
secondary current 二次电流
secondary ion mass spectroscopy (SIMS) 二次离子质谱学
secondary winding 二次绕组,次级绕组
sectioning device (电)分段器
section insulation 段间绝缘,分段绝缘
sector 扇形叶片;扇区,扇段
sediment 沉积物;沉积
seepage control installation 防渗设施
segregation 分离;离析
self-aligning bearing 自位轴承
self-baking electrode 自焙电极
self-checking 自检
self-closing 自动关闭
self-commutation 自换相
self-cooled type 自冷式
self-diagnosis of fault 事故自诊断
self-excitation 自励
self-extinguishing 自熄
self-extinguishing fault 自熄弧故障
self-feeder 自动给料机
self-heating 自(加)热
self-holding 自保持
self-inspection 自检,自查
self-resetting 自复位
self-restoring insulation 自恢复绝缘
self-starter 自(动)起动器
self-triggering 自触发
selsyn 自整角机;自同步机
SEM (scanning electron microscope) 扫描电子显微镜
semi-anechoic chamber 半电波暗室
semi-automatic logging instrument 半自动测井仪
semi-automatic testing machine 半自动试验机
semi-automation 半自动化
semiconducting specimen stage 半导体样品台
semiconductor strain gauge 半导体应变计
semiconductor switch device 半导体开关装置
semidigital readout 半数字读出
semi-graphic panel 半模拟盘,半模拟屏

semi-physical simulation 半实物仿真
sense circuit 读出电路
sensibility 灵敏度
sensor 传感器,敏感器,敏感元件
sensor with bare wire sensitive element 裸丝敏感元件传感器
sensor with external convection sensitive element 外对流敏感元件传感器
sensor with internal convection sensitive element 内对流敏感元件传感器
sensor with supporter catalyst filled element 载体催化元件传感器
separate excitation 他励
separator （粗粉）分离器；隔板,隔离层
sequence chart 序列表图
sequential auto-spectrometer 顺序式自动光谱仪
sequential control 顺控,顺序控制
sequential controller 顺序控制器
sequential control station 顺序控制站
sequential control system 顺序控制系统,顺控系统
sequential decomposition 顺序分解
sequential least squares estimation 序贯最小二乘估计
sequential optimization 顺序优化
sequential program 顺序程序
sequential topology 顺序拓扑
sequential X-ray spectrometer 扫描 X 射线光谱仪
serial access 串行存取
serial computer 串行计算机
serial operation 串行操作
serial processing 串行处理
serial transmission 串行传输
series capacitor 串联电容器
series circuit 串联电路
series connection 串联

series inductor 串联电感器
series-mode rejection 串模抑制
series-mode rejection ratio (SMRR) 串模抑制比
series-mode signal 串模信号
series-mode voltage 串模电压
series operation 串联运行
series output system 串行输出制
series-parallel circuit 串并联电路
series-parallel output system 串并行输出制
series resistor 串联电阻器
serpentine tube 蛇形管,螺旋管
serviceability 运行能力；可服务性；适用性
service condition 使用条件,运行条件
servoamplifier 伺服放大器
servomotor 伺服电(动)机
servomotor actuator 伺服电动执行器
seston 悬浮物
set point 设定点
set screw 紧定螺钉
setting 整定
settler 澄清器
settling time 建立时间
sewage 污水
sewer 排水管渠,下水道
sewerage 排水工程；污水
shackle insulator 蝶式绝缘子
shade 灯罩
shaft 轴；竖井,通风井
shaft current 轴电流
shaft spillway 井式溢洪道
shaft-voltage test 轴电压试验
shaft work 轴功
shaker hearth furnace 震底式炉
shale 页岩
shank 刀柄,钻头柄
shaped conductor 异形导线
shear 剪切；剪切机
sheath 护套
sheathing 护套
sheave 滑车轮,槽轮

sheet 板材,薄板
sheeting 护墙板;压片
sheet resistor 片电阻
shelf life 贮存寿命
shell boiler 火管锅炉,锅壳式锅炉
sheradizing 渗锌
shield 防护屏;接地屏蔽
shield method 盾构法
shifter 移动装置;移位器
shifting 移位
shim 垫片,填隙片
shim rod 补偿棒
ship lift 升船机
ship lock 船闸
shock excitation 冲击激励
shockproof packaging 防震包装
shock response spectrum 冲击响应谱
shock sensor 冲击传感器
shock test 冲击试验
shock testing machine 冲击(试验)台
shock transducer 冲击传感器
shock vibration 冲击振动
shoot 斜道,滑道
shop-assembled fabricated piece (SAFP) 车间组装制作件
shore 支撑柱
short circuit 短路
short-circuit between strands 股线间短路
short-circuit capacity 短路容量
short-circuit commutator test 短路换向器试验
short-circuit making 短路接通
short-circuit operation 短路运行
short-circuit-proof transformer 耐短路变压器,短路保护变压器
short-circuit ratio 短路比
short-circuit turn 短路匝
short-line fault (SLF) 近区故障
shotcrete 喷射混凝土
shoulder load 中间负荷
shredder 粉碎机,切碎机
shrinkage 收缩;收缩率

shrink fit 热套配合,冷缩配合
shrink-on disc rotor 套装转子
shroud 围带
shuttering 背板
shuttle 滑闸,闸门;振荡传输机;梭
SI[1] (special information) 专用信息
SI[2] (special instruction) 专用说明书
sieve 筛子;筛
signal 信号
signal amplitude sequencing control 信号幅值顺序控制
signal cable 信号电缆
signal circuit 信号电路
signal converter 信号转换器
signal detection and estimation 信号检测和估计
signal duration 信号持续时间
signal flow diagram 信号流(程)图
signal isolation 信号隔离
signal level 信号电平
signal processing 信号处理
signal processing system 信号处理系统
signal repeater 信号复示器
signal selector 信号选择器
signal-to-noise ratio 信噪比
SIL (switching impulse level) 操作冲击水平
silencer 消声器
silica 二氧化硅
silicate 硅酸盐
silicon controlled rectifier (SCR) 可控硅整流器
silicon integration 硅集成
silo 筒仓
silt 淤泥;粉土,粉砂
simplex lap winding 单叠绕组
simplex wave winding 单波绕组
SIMS (secondary ion mass spectroscopy) 二次离子质谱学
simulated data 仿真数据
simulated interrupt 仿真中断
simulating strain 模拟应变

simulating strain device 模拟应变装置
simulation 仿真,模拟
simulation algorithm 仿真算法
simulation algorithm library 仿真算法库
simulation block diagram 仿真(方)框图
simulation centre 仿真中心
simulation clock 仿真时钟
simulation database 仿真数据库
simulation environment 仿真环境
simulation equipment 仿真设备
simulation evaluation 仿真评价
simulation experiment 仿真实验
simulation experiment mode library 仿真实验模式库
simulation expert system 仿真专家系统
simulation graphic library 仿真图形库
simulation information library 仿真信息库
simulation job 仿真作业
simulation knowledge base 仿真知识库
simulation laboratory 仿真实验室
simulation language 仿真语言
simulation methodology 仿真方法学
simulation model 仿真模型
simulation model library 仿真模型库
simulation process 仿真过程
simulation process time 仿真过程时间
simulation program 仿真程序
simulation result 仿真结果
simulation run 仿真运行
simulation software 仿真软件
simulation support system 仿真支持系统
simulation system 仿真系统
simulation technique 仿真技术
simulation type 仿真类型
simulation velocity 仿真速度
simulation work station 仿真工作站
simulator 仿真器
simultaneous comparison method 同时比较法
simultaneous technique 同时联用技术,同时并用技术
sine 正弦
sine-cosine resolver 正余弦旋转变压器
sine function 正弦函数
sine winding 正弦绕组
single-cylinder machine 单缸汽轮机
single-phase circuit 单相电路
single-phase system 单相系统
single-pole, double throw (SPDT) 单刀双掷
single-pole, single-throw (SPST) 单刀单掷
sink 汇(点)
sinusoidal variation 正弦变化
siphon 虹吸管
SI unit 国际单位,SI 单位
skewback 拱座
skew factor 斜槽因数
skid hearth furnace 滑底式炉
skid wire 滑线
skim 表层物;撇渣
skin casing 外护套
slab foundation 板式基础
slackening of iron core 铁芯松弛
slag 炉渣,熔渣
slaked lime 熟石灰
slaking 熟化(石灰)
slate 板岩
sleeve 插套,套筒
slewing 旋转
SLF (short-line fault) 近区故障
slider 滑动块;滑块,滑板
slide valve 滑阀
sliding 滑动,滑移
sliding contact 滑动接触;滑动触头

slime 煤泥,黏垢
slip ring 集电环,滑环
slot 槽,线槽,插槽
slot wedge 槽楔
slow-front overvoltage 缓波前过电压
sludge 油泥;泥渣,沉渣
sluice 水闸;人工泄水道
slurry 泥浆,浆体
smart battery 灵巧电池
smelter 熔炉;冶炼厂
SMMR (scanning microwave multiband radiometer) 多波段扫描微波辐射仪
SMPS (switch-mode power supply) 开关式电源;开关式供电
SMRR (series-mode rejection ratio) 串模抑制比
snap-on contact 瞬接触点
snow load 雪载
snubber 电压尖脉冲缓冲电路;缓冲器,减震器
socket 插座;灯座;帽窝
socket outlet 插座
sodium iodide 碘化钠
softener 软化器;软化剂
softening 软化
soft magnetic material 软磁材料
soil erosion 土壤侵蚀
solar energy 太阳能
solar photovoltaic conversion 太阳能光电转换
solder 焊料
soldering 钎焊
solid 固体
solid conductor 实心导体,实心导线
solidification 凝固,固化
solid radioactive waste 固体放射性废物
solid solution semiconductor 固溶体半导体
solubility 溶解度,可溶性
solute 溶质,溶解物
solution 溶液

solvend 可溶物质
solvent 溶剂;有溶力的
sonar 声呐;声呐定位法;声波定位仪
sonic speed 声速,音速
soot 烟粒,煤烟
soot blower 吹灰器
soundness 可靠性,有效性
sound pressure level (SPL) 声压级
sour gas 酸气
spacer 隔离物;隔叶块;间隔棒
spacing 结构高度;间隔,间距
spade 铲
span 跨度;档距;量程
spanner 扳手
spar 晶石;亮晶
spark emission 散发火花
spark ignition 火花点火
sparkless operation 无火花运行
spark plug 火花塞
spark plug suppressor 火花塞抑制器
spark-proof 防火花的
spark tester 火花检测器
SPC (stored-program control) 存储程序控制
SPDT (single pole, double throw) 单刀双掷
special fastener 特殊紧固件
special information (SI) 专用信息
special instruction (SI) 专用说明书
special trochoid 特殊余摆线
specific speed 比转速
specific volume 比容,比体积
spectral background 光谱背景
spectral bandwidth 谱带宽度
spectral characteristic curve 光谱特性曲线
spectral density (频)谱密度
spectral distribution curve 光谱分布曲线
spectral distribution of energy 光谱能量分布

spectral emissivity 光谱发射率
spectral half width 光谱半宽度
spectral line (光)谱线
spectral position 光谱位置
spectral radiance 光谱辐射亮度
spectral range 光谱范围
spectral resolution 波长分辨率；谱分解
spectral slit width 光谱狭缝宽度
spectrochemical analysis 光谱化学分析
spectrofluorophotometer 荧光分光光度计
spectrograph 摄谱仪
spectrometer 光谱仪
spectrophotometer 分光光度计
spectrophotometric titration 分光光度滴定法
spectroradiometer 光谱辐射计
spectroscopy 光谱学，能谱法
spectrum 谱，光谱；频谱
spectrum analyzer 频谱分析仪
speed governor 调速器
speed regulator 调速器
speed-sensitive output voltage 速敏输出电压
sphericity 球形度，圆球度
spider 转子支架；星形轮；多脚架
spike 尖峰信号
spillway 溢洪道
spindle 主轴；锭子
spinning reserve 旋转备用
spiral resistor 螺旋形电阻体
SPL (sound pressure level) 声压级
spline 花键；样条
split brush 分瓣电刷
splitter 分路器，分流器
splitting 解列；分流
spool 线轴
spool insulator 线轴式绝缘体
spot welding 点焊
spout 出料槽
sprayer 喷雾器
spread factor 分布因数
spring contact 弹性触头
spring pressure 弹簧压强

sprocket 链轮
SPST (single-pole, single-throw) 单刀单掷
SPT (standard penetration test) 标准贯入试验
spun glass 玻璃纤维
squeezing thickness 排挤厚度
squirrel-cage rotor 鼠笼式转子
SS^1 (stainless steel) 不锈钢
SS^2 (suspended solid) 悬浮物，悬浮固体
SSC (submerged scraper conveyor) 水浸式刮板捞渣机
S-shaped trap S形曲颈管
stabilization 稳定(化)，稳定作用
stabilizer 稳定器；稳定剂
stable ignition 稳定着火
stack 烟囱；炉身；垛；栈
stacked memory (堆)栈存储器
stacker 堆料机，堆叠器
stacker-reclaimer 堆取料机
stack gas 烟气
stadia 视距仪
stage (汽轮机)级
staggering peak load 错峰
stained glass 有色玻璃
stainless steel (SS) 不锈钢
stamping 冲压；冲压件
standard annealed copper 标准软铜
standard coal 标准煤
standardization 标准化
standard penetration test (SPT) 标准贯入试验
standing loss 保温损耗
Stanton number 斯坦顿数
star connection 星形连接
star quad 星形四线组
star quadding machine 星绞机
start-up 启动
static accuracy 静态精确度
static balance 静平衡
static calibration 静态校准
static characteristic curve 静态特性曲线
static characteristics 静态特性

static decoupling 静态解耦
static display image 静态显示图像
static-dynamic strain indicator 静动态应变仪
static/dynamic universal testing machine 动静万能试验机
static electricity 静电
static friction 静摩擦
static gauging method 静态(容积)测量法
static load 静态负荷
static load compensating device 静态负荷补偿装置
static mass spectrometer 静态质谱仪
static measurement 静态测量
static model 静态模型
static precision 静态精密度
static pressure 静压
static pressure sensor 静态压力传感器
static pressure transducer 静态压力传感器
static relay 静态继电器
statics 静力学;静态
static SIMS 静态二次离子质谱法
static spring gravimeter 静力弹簧重力仪
static standard strain device 静态标准应变装置
static strain 静应变
static strain indicator 静态应变仪
static synchronizing torque 静态整步转矩
static thermal measuring instrument 静态热测量仪表
static thermal technique 静态热技术
static torque 静转矩,静扭矩
static transformer 静电变压器
static unbalance 静不平衡
static weighing method 静态称重法
static weight-hoist 静态挂码
stator 定子
stator winding machine 定子绕嵌机
stay 拉线;静子
steady state 稳态
steam 蒸汽,水蒸气
steam bending stress 蒸汽弯曲应力
steam chest (蒸)汽室,(汽轮机)进汽箱
steamer 气锅
steam generator 蒸汽锅炉
steam jet 蒸汽喷射流
steam-jet air ejector 射汽抽气器
steam lead 蒸汽导管,蒸汽管道
steam line blowing 蒸汽管路吹扫
steam power plant 蒸汽(动力)发电厂
steam receiver 集汽管
steam turbine 汽轮机
steam washing 蒸汽清洗
steel 钢
steel wire gauge 钢丝规
steepness of voltage collapse 电压骤降陡度
stellite (alloy) 司太立合金
STEM (scanning transmission electron microscope) 扫描透射电子显微镜
stem 芯柱
step-down transformer 降压变压器
stepped labyrinth gland 阶梯式迷宫式汽封
stepping motor 步进电动机
stepping switch 步进式开关
step pitch 步距
step-up transformer 升压变压器
sterilizer 消毒器
stiffener 加劲杆,加强条,加劲肋
stiffness 硬挺度,劲度,刚度,刚性
stirrer 搅拌器
stirring 搅拌

STM (scanning tunnelling microscope) 扫描隧道显微镜
stoichiometry 化学计量
stoker 加煤机
stop 限位器,挡块
stop joint 塞止接头
stop valve 断流阀,截止阀
storage capacity 库容
storage cell 蓄电池
storage medium 存储介质
storage oscilloscope 存储示波器
storage tube 存储管
storativity 储水系数
stored-program control (SPC) 存储程序控制
straightener 矫直机,拉直机
strain 应变;张力;过滤
strainer 滤网,粗滤器,过滤器;拉紧装置
strain insulator 拉紧绝缘子
strainmeter 应变仪
stranded conductor 绞线,绞合导线,绞合导体
strangler 节流门
stratum 地层
stray loss 杂散损耗
streamline 流线
strength 强度
stress concentration 应力集中
stress corrosion 应力腐蚀
stress gauge 应力计
stretcher 拉伸机,拉紧机
stretching 拉伸,张拉
striker 撞击器
striker fuse 撞击熔断器
strip-on-edge winding machine 扁绕机
strobe pulse 选通脉冲
stroke 冲程,行程
Strouhal number 施特鲁哈尔数
structural controllability 结构可控性,结构能控性
structural coordination 结构协调
structural decomposition 结构分解
structural formula 结构(公)式
structural observability 结构可观测性,结构能观测性
structural passability 结构可通性,结构能通性
structured program 结构化程序
structured programming 结构化程序设计
structured programming language 结构化程序设计语言
structure model 结构模型
strut insulator 棒式绝缘子
stub 短截线
stub shaft 加伸轴
stucco 粉饰灰泥
stud 双头螺栓,大头钉,栓钉
stuffing 填料
sub-assembly 部件,装配件,分部件,分组件
sub-bituminous coal 次烟煤
sub-conductor 子导线
subcooling 过冷却,低温冷却
subcritical 亚临界
subcritical pressure boiler 亚临界压力锅炉
subcritical pressure turbine 亚临界压力汽轮机
subdrainage 地下排水系统
sublimation 升华
submerged arc furnace 埋弧炉
submerged arc heating 埋弧加热
submerged scraper conveyor (SSC) 水浸式刮板捞渣机
subscript 下标
subsidence 沉陷,沉降,土壤下降
substation 变电站,变电所
suction head 吸入压头
suction height 吸上高度,吸入高度
sulphur 硫
sulphurization 硫化
sump 集油槽;集水坑,污水坑
superconductivity 超导(电)性
supercritical plant 超临界机组
superfluid 超流体
superheating 过热
superheater 过热器

supply duct 送风管道
supply outlet 送风风口
suppressant 抑制剂
suppressor 抑制剂;消除器
suppression 抑制
suppression capacitor 抑制电容器
suppression component 抑制元件
suppressive wiring technique 抑制布线技术
surface-type heater 表面式加热器
surge tank 调压井
surge tower 调压塔
susceptance relay 电纳继电器
susceptibility 磁化率;敏感度
susceptivity 敏感度
suspended boiler structure 悬吊式锅炉构架
suspended solid (SS) 悬浮物,悬浮固体
suspender 吊杆,吊索
suspension 悬浮;悬浮液;挂起
suspension insulator 悬式绝缘子
sustainability 可持续性
SV (safety valve) 安全阀
sweating 析出
sweet gas 甜气,净气
swelling of case 外壳鼓胀
swing curve 摇摆曲线
swirler 旋流器
switch 开关;切换;开合,通断
switch arm 开关臂
switchboard 配电盘
switch-fuse 带熔断器开关,熔丝开关
switchgear 配电装置;开关设备
switching impulse level (SIL) 操作冲击水平
switching impulse test 操作冲击试验
switching overvoltage 操作过电压
switch-mode power supply (SMPS) 开关式电源;开关式供电

switchyard 开关场
symmetry 对称性
synchronism 同步
synchronizer 同步器,同步机,同步装置
synchronizing 整步
synchronizing reactor 整步电抗器
synchronizing winding 整步绕组
synchronous control 同步控制
synchroscope 同步指示器
syncline 向斜
syphon 虹吸管
system 系统,体系
system aggregation 系统集结
system analysis 系统分析
system approach 系统方法
system architecture 体系结构
system assessment 系统评价
systematic error 系统误差
systematic uncertainty 系统不确定度
systematology 系统学
system decomposition 系统分解
system design description (SDD) 系统设计说明
system deviation 系统偏差
system diagnosis 系统诊断
system environment 系统环境
system evaluation 系统评价
system homomorphism 系统同态
system identification 系统辨识
system impedance ratio 系统阻抗比
system interrupt 系统中断
system isomorphism 系统同构
system maintainability 系统可维护性
system management 系统管理
system matrix 系统矩阵
system model 系统模型
system modelling 系统建模,系统模型化
system of units (测量)单位制
system optimization 系统(最)优化
system parameter 系统参数

system performance test 系统性能试验
system planning 系统规划
system production time 系统生产时间
system reliability 系统可靠性
system resource 系统资源
systems engineering 系统工程
system sensitivity 系统灵敏度
system simulation 系统仿真,系统模拟
system software 系统软件
system state 系统状态
system statistical analysis 系统统计分析
system test time 系统测试时间
system theory 系统理论
system with effectively earthed neutral 中性点有效接地系统

table-driven compiler 表驱动编译程序
tablet 片剂;输入板
tablet coating 压片涂层
tabular display 表格显示(器)
tabular sequence control 列表时序控制
tabular value 表(列)值
tabular values of Bessel function 贝塞尔函数表值
tabulated solution 表解
tabulating card 制表卡片
tabulating machine 制表机
tabulating system 制表系统
tabulation 制表,列表
tabulator 制表机
TAC (time-to-amplitude converter) 时间-幅度变换器
tachogenerator 测速发电机
tachometer 转速表
tachometer generator 测速发电机
tachometer sender 测速发送机
tachometer stabilized system 转速表稳定系统
tachometric relay 转速继电器
tachometry 转速测定法
tack 平头钉,图钉
tackle 滑车;索具
tackle block 滑轮
tack welding 定位焊,平头焊
TACT (transistor and component tester) 晶体管及元件测试仪
tactron 冷阴极充气管
tact system 流水作业
TADP (terminal area distribution processing) 终端区域分布处理
tag wire 终端线

tail channel (水电站的)尾水渠
tail effect 尾端效应
tail end 尾端
tail-end booster 线路末端升压器
tail gate 尾水闸门
tail heating surface 尾部受热面
tailing peak 拖尾峰
tailing(s) 尾矿,尾渣,残渣
tail-of-wave impulse test voltage 冲击[闪络]试验波尾电压
tailor-made motor 定制电动机
tailrace (涡轮)放水道,尾水渠
tailrace surge chamber 尾水调压室
tailrace tunnel 尾水隧洞
tail stream 尾料流
tailwater gallery 尾水廊道
takeover current 交接电流
talk-back circuit 内部对讲电路,回话电路
tall stack source 高烟囱(污染)源
tally order 总结指令,结算指令
tamped backfill 夯实回填土
tamped cinder 夯实炉渣
tamper 夯,打夯机
T & D (transmission and distribution) 输电与配电,输配电
tandem arrangement 纵列配置,串联配置
tandem blade 串列式叶片
tandem-completing trunk 转接中继线
tandem compound turbine 单轴(系)汽轮机,纵联复式汽轮机
tandem compound turbogenerator 纵联复式汽轮发电机

tandem connection 串联
tandem control 串联控制
tandem double-flow turbine 单轴双排汽汽轮机
tandem engine 串列式发动机
tandem generator 串联起电器;串联发电机
tandem knife switch 串联刀开关
tandem mirror experiment 串级磁镜试验
tandem mirror fusion-fission hybrid 串级磁镜聚变裂变混合堆
tandem mirror reactor 串级磁镜反应堆
tandem motor 双电枢共轴电动机,级联电动机
tandem office 中继局,转接局
tandem selection 中继选择,转接选择
tandem sequence submerged arc welding 纵列多丝埋弧焊
tandem switch 汇接中继线键
tandem transistor 双晶体三极管
tandem turbine 单轴汽轮机
tandem type pump 串联式水泵
tandem winding coil 纵列绕组线圈
tangency condition 相切条件
tangent 切线;正切
tangent galvanometer 正切电流计
tangential acceleration 切向加速度
tangential admission 切向进汽
tangential approximation method 切线逼近法
tangential beam tube 切向束射管
tangential blade spacing 切向叶片间距
tangential burner 角式燃烧器
tangential component 切向分量
tangential corner firing 切向四角燃烧
tangential couple 切向力偶
tangential fan 横流扇,贯流式通风机
tangential firing 切向燃烧
tangential flow turbine 切击式水轮机,双击式水轮机
tangential force 切向力
tangential force coefficient 切向力系数
tangential gradient probe 切向梯度探头
tangential keyway 切向键槽
tangential line 切线
tangentially fired boiler 切向燃烧锅炉
tangentially fired combustion chamber 切向燃烧炉膛
tangentially fired furnace 切向燃烧炉膛
tangential nozzle 切向接管
tangential out-of-phase 切向异相位
tangential partial turbine 切向非整周进水式水轮机
tangential shearing stress 切向剪应力
tangential strain 切向应变
tangential stress 切向应力
tangential thrust 切向推力
tangential turbine 切击式水轮机
tangential velocity 切向速度
tangential wave 切向波
tangent line 切线
tangent modulus matrix 正切模量矩阵
tangent plane 切面
tangent scale 正切标度;正切尺
tangent stiffness matrix 切线刚度矩阵
tangent tower 直线塔
tangent tube construction 密排管结构(无间隙)
tangent tube wall 密排管水冷壁(无间隙)
tangent welding 多级平行弧焊
tank bottom heater 箱底加热器
tank capacitor 槽路电容器,振荡回路电容器

tank circuit 谐振电路,槽路
tank cooler 油箱冷却器
tank level indicator 箱液面指示器
tank reactor 槽型反应器
tank retention 桶内贮存
tank transformer 油冷变压器,油箱式变压器
tank-type circuit-breaker 箱式断路器
tank-type reactor 箱式(反应)堆
tank voltage 槽电压
tantalum capacitor 钽电容器
tantalum rectifier 钽整流器
tap 抽头,分接头;丝锥
tap adjuster 分接头调整器
tap-change operation 分接变换操作,分接头换接
tap changer 抽头换接开关,分接头换接装置
tap-change transformer 抽头切换变压器
tap changing 抽头切换,分接头切换
tap-changing transformer 抽头切换变压器
tap drill 螺孔钻,螺丝底孔钻
tape armour 钢带铠装
tape bootstrap routine 带引导例程
tape comparator 带比较器
tape control unit 带控装置
tape covering 带绝缘层
taped insulation 带绕绝缘
tape drum 带鼓
tape identification unit 带标识器
tape insulated coil 带绝缘线圈
tape insulation 带绝缘
tape limit 带限(制)
tape-loop storage 带环存储器
tape mechanism 卷带机构
tape operating system (TOS) (磁)带操作系统
tape parity 带奇偶(校验)
tape plotting system 磁带绘图系统
tape preparation unit 带准备装置
tape punch 纸带穿孔(机)
taper cone 圆锥
tape reader 纸带读入机
tapered aeration 渐减曝气
tapered blade 渐变截面叶片,锥形叶片
tapered cascade 锥形级联
tapered plug valve 锥塞阀
tapered potentiometer 递变电阻分压器
tapered slide valve 锥形滑阀
tapered tube rotameter 带锥形管柱的转子流量计
taper-loaded cable 端部负载渐减电缆,锥形负载电缆
taper pin 圆锥销
taper reamer 圆锥铰刀
taper wedge 斜楔
tape skip 越带指令
tape transmitter-distributer 带发送分配器
tape-wound core 带绕磁芯
tap-field motor 分段励磁绕组电动机
taping machine 包带机
tapped air-gap reactor 分段气隙电抗器
tapped coil 多(接)头线圈,抽头线圈
tapped control 分接头控制
tapped delay line 抽头延迟线
tapped horsepower rating 分级功率(调速直流电动机)
tapped induction coil 抽头感应线圈,多头感应线圈
tapped line 支线,抽头线;多抽头线路
tapped stator winding 抽头定子绕组
tapped transformer 抽头式变压器
tapped tuned circuit 抽头调谐电路
tapped winding 多抽头绕组
tappet gear 挺杆传动装置

tappet switch 制动开关(带导轮的开关)
tapping 分接
tapping adjuster (变压器的)分接开关调节器
tapping attachment 攻丝装置,攻丝夹头
tapping contactor 分接接触器
tapping current 分接电流
tapping factor 分接因数
tapping point 分接头;(汽轮机的)抽汽口;分支点
tapping power 分接容量
tapping quantities 分接参数;分接量
tapping range 分接范围
tapping ratio 抽头匝数比
tapping screw 自攻螺钉
tapping screw thread 自攻螺纹
tapping step 分接级
tapping switch 分接开关
tapping voltage 分接电压
tap position indicator 分接头位置指示器
tap position information 分接头位置信息
tap-selector 分接头选择器,分接开关
tap-selector drum switch 鼓形分接头选择开关
tap switch 分线开关
tap voltage 抽头电压,分接头电压
TARE (transistor analysis recording equipment) 晶体管分析记录设备
tare-load ratio 自重-载重比
tar epoxy paint 柏油环氧漆
tare (weight) 皮重
target 靶;目标
target acquisition 目标捕获
target area 目标区;靶区
target burn-up 目标燃耗
target computer 特定程序计算机,结果程序计算机
target cut-off voltage 靶截止电压(摄像管)
target flowmeter 靶式流量计
target irradiation 目标辐照
target nucleus 靶核
target variable 目标变量
tarnish 失泽物
tarring number 焦(油)化值
tarring value 焦(油)化值
tar sand 焦油砂
tartaric acid 酒石酸,2,3-二羟基丁二酸($C_4H_6O_6$)
tartaric acid solution 酒石酸溶液
TAS (test and set) 测试与置位
task control block (TCB) 任务控制块
task descriptor 任务描述符
task dispatcher 任务分派程序
task execution memory 任务执行存储器
T-attenuator T型衰减器
tautology 重言式
Taylor build-up factor 泰勒积累因子
Taylor connection 泰勒接线法
Taylor expansion 泰勒展开
Taylor formula 泰勒公式
Taylor series 泰勒级数
Taylor standard screen 泰勒标准筛
TB[1] (terminal board) 端子板,端子排
TB[2] (thermal balance) 热(量)平衡
TBC (turbine bypass control) 汽轮机旁路控制
TBE (time-base error) 时基误差
TBFP (turbo boiler feed pump) 汽动锅炉给水泵
TBO (time between overhauls) 大修间隔
TBP (turbine bypass) 汽轮机旁路
TC[1] (temperature compensation) 温度补偿
TC[2] (temperature controller) 温度控制器

TC³ (thermal couple; thermocouple) 热电偶,温差电偶
TC⁴ (transmission controller) 传输控制器
TC⁵ (turbine-compressor) 汽轮机-压气机组
TCB (task control block) 任务控制块
TCE (transmission control element) 传输控制元件
T-circuit T形电路
TCL (transistor-coupled logic) 晶体管耦合逻辑
TCM (terminal-to-computer multiplexer) 终端-计算机多路转换器
TCO (trunk cut-off) 干线切断
TCP (transmission control protocol) 传输控制协议
TCS¹ (terminal control system) 终端控制系统
TCS² (thermal control system) 热控制系统
TCSL (transistor current steering logic) 晶体管电流导引逻辑
TCTL (transistor-coupled transistor logic) 晶体管耦合晶体管逻辑
TCU (temperature control unit) 温度控制装置
TD¹ (technical data) 技术数据
TD² (temperature difference) 温差
TD³ (terminal difference) 端差
TD⁴ (time delay) 时延,时间延迟
TD⁵ (transmitter distributor) 发射机分配器
TD⁶ (tunnel diode) 隧道二极管
TDF (two degrees of freedom) 二自由度
TDM (time-division multiplexing) 时分复用
TDP (teledata processing) 远程数据处理
TDS¹ (time-division switch) 时分开关
TDS² (total dissolved solids) 总含盐量,总溶解固体
TE¹ (test equipment) 测试设备
TE² (thermal efficiency) 热效率
TE³ (trailing edge) 后缘
teaser coil 梯塞线圈,调节变压比的线圈
teaser winding 梯塞绕组,可调绕组,辅助绕组
TEC (thermal expansion coefficient) 热膨胀系数
technetron 场调管,场效应高能晶体管
technical coefficient 技术系数
technical data (TD) 技术数据
technical design 技术设计
technical inspection report 技术检验报告
teetered hub 翘板型轮毂
telecommunication cable 通信电缆
telecontrol 远动,遥控
telecounting 远程计算
teledata processing (TDP) 远程数据处理
Tellegen's theorem 特勒根定理
temperature coefficient 温度系数
temperature-compensated electric machine 温度补偿电机
temperature compensation (TC) 温度补偿
temperature controller (TC) 温度控制器
temperature control unit (TCU) 温度控制装置
temperature difference (TD) 温差
temperature limiter 限温器
temperature-resistance curve 温度-电阻曲线
temporary overvoltage 暂态过电压
tensor permeability 张量磁导率
tensor susceptibility 张量磁化率
terminal area distribution processing (TADP) 终端区域分布处理

terminal board (TB) 端子板,端子排
terminal box 终端盒;接线盒
terminal control system (TCS) 终端控制系统
terminal difference (TD) 端差
terminal interface module (TIM) 终端接口模件
terminal-to-computer multiplexer (TCM) 终端-计算机多路转换器
terminal tower 终端塔
terminal wire 端线
tertiary air 三次风
tertiary winding 三次绕组,第三绕组
test and set (TAS) 测试与置位
test equipment (TE) 测试设备
test run (TR) (测)试运行
theoretical numerical aperture 理论数值孔径
thermal balance (TB) 热(量)平衡
thermal blockage 热挂
thermal capacitance 热容
thermal control system (TCS) 热控制系统
thermal couple (TC) 热电偶,温差电偶
thermal derating factor 热降额因数
thermal deviation 热偏差
thermal diffusivity 热扩散率
thermal efficiency (TE) 热效率
thermal endurance profile 耐热概貌
thermal expansion coefficient (TEC) 热膨胀系数
thermal fatigue 热疲劳
thermal life 热寿命
thermal neutron 热中子
thermal loss 热损失
thermal radiator 热辐射体
thermal resistance 热阻
thermal runaway 热崩溃,热失控
thermal stability test 热稳定试验
thermal time-delay switch 热延时开关
thermal transmission coefficient 传热系数
thermionic emission 热离子发射
thermistor 热敏电阻(器)
thermo-bimetal 热双金属
thermocouple (TC) 热电偶,温差电偶
thermodynamic cycle 热力学循环
thermodynamic system 热力学系统
thermo-electric material 电热材料
thermoelectric potential 热电势
thermo-electric traction 热-电牵引
thermogravimetry 热失重法
thermoset plastics 热固性塑料
thermosiphon 热虹吸器
thermostat 恒温器
thermostatic switch 定温开关
Thevenin theorem 戴维南定理
Thomson bridge 汤姆孙电桥
three-phase circuit 三相电路
three-phase three-limb core 三相三柱式铁芯
threshold current 阈电流
threshold ratio 阈值比
threshold voltage 阈电压
throat depth 电极臂伸出长度
throat gap 电极臂间距
throttling loss 节流损失
throughway bushing 穿通式套管
throw 摆度
thrust bearing 推力轴承
thrust collar 推力盘
thyristor module 晶闸管模块
ticker 断续装置;自动收报机;蜂音器
tickler (coil) 反馈线圈,回授线圈
tidal current 潮流
tidal datum 潮高基准面

tidal energy 潮汐能
tidal power 潮汐能,潮汐动力
tidal power generation 潮汐发电
tidal power generator 潮汐发电机
tidal power station 潮汐电站
tidal wave 潮(汐)波
tide gauge 验潮仪
tide-meter 验潮仪
tide-motor 潮汐发动机
tide pole 验潮杆
tie bolt 拉紧螺栓,承拉螺栓
tie-down insulator 悬垂绝缘子
tie-in transformer 联络变压器
tie jumper 连接跳线
tie line 联络线;结线
tie point 连接点;(电力系统)联络点
tie-rod stator frame 拉杆式定子框架
tight-closing isolating damper 密闭隔离挡板
tight coupling 紧耦合,强耦合
tight lattice 稠密栅格
tight-lock coupler 密锁自动耦合器
tight-packed lattice 稠密栅格
tight tube fibre 紧套光纤
tilter 焊接翻转机;倾卸装置
tilting bar 翻转炉排片
tilting burner 摆动(式)燃烧器
tilting cradle 倾炉架
tilting furnace 倾倒式炉
tilting manometer 倾斜压力计
tilting micromanometer 倾斜式微压计
tilting moment 倾覆力矩
tilting nozzle 摆动喷嘴
tilting-pad bearing 可倾瓦块轴承,斜垫轴承
tilting-ring tachometer 摆环式转速计
tilting trestle 翻料架
tilt motor 俯仰操纵电动机
TIM (terminal interface module) 终端接口模块
time-anticoincidence circuit 时间反符合电路
time-average fringe pattern 时间平均条纹图样
time-average hologram 时间平均全息图
time-average holography 时间平均全息术
time-average interferometry 时间平均干涉测量术
time axis 时间轴
time base 时基;时间坐标
time-base circuit 时基电路
time-base error (TBE) 时基误差
time-base generator 时基发生器
time between failures 无故障工作时间
time between overhauls (TBO) 大修间隔
time bias 时偏;时间补偿(电力系统)
time-code generator 时间码生成器
time coherence 时间相干性
time-coincidence circuit 时间符合电路
time-consolidation curve 时间-固结曲线
time constant 时间常数
time controller 自动定时仪
time correlation 时间关联
time-current threshold 时间电流阈值
time cut-off 定时断流器
time cut-out 定时断路器
time delay (TD) 时延,时间延迟
time-delay circuit 延时电路
time-delay contactor relay 延时继电接触器
time-delay push button 延时复位按钮
time-delay relay 延时继电器,缓动继电器
time-delay servo (system) 时延伺服系统,时延随动系统
time-delay stopping relay 延时停机继电器

time-delay undervoltage relay 延时欠电压继电器
time dependence 时间相关
time derivative 时间导数
time discriminator 鉴时器
time distance relay 时间距离继电器
time division 时分,时间划分
time-division multiplexing (TDM) 时分复用
time-division multiplier 时分多路乘法器
time-division switch (TDS) 时分开关
time-division switching 时分(制)交换
time-division system 时分制
timed magneto 定时磁电机
time-domain electromagnetics 时域电磁学
time-domain matrix 时域矩阵
timed pulse 时控脉冲,定时脉冲
timed relay 定时继电器
time edge effect 时间边缘效应
time-element relay 时限元件继电器,延时继电器
time-frequency duality 时间-频率对偶性
time harmonic (电压和电流波形的)时间谐波
time history analysis method 时程分析法
time integral 时间积分
time-interval indicator 时间间隔指示器
time-invariant regulator 定常调整器,时间恒定调整器
time-invariant system 定常系统,时间恒定系统,非时变系统
time-lag effect 时滞效应
time-lag fuse 延时熔断器
time-lag relay 延时继电器,时限继电器
time-limit attachment 时限附件
time-limit breaker 时限(自动)断路器
time-limit relay 时限继电器
time lock 时间同步,时间锁定
time modulation 时间调制
time monitor 时间监视器
time overcurrent 时限过电流
time phase 时间相位
time-phase angle 时间相角
time phasor 时间相量
time pulse 时间脉冲
time pulse distributor 时间脉冲分配器
time pulse relay 定时脉冲继电器
time quadrature 正交时间相移,正交时间相位差
timer 计时器,定时器
time rate 时(间)变率
time rating 额定(工作)时间;(连续或断续)短时运行功率
time recorder 时间记录器
time relay 时间继电器,定时继电器
time release 延时释放(器)
time scale 时间标度;时间比例;时间量程
time-scale factor 时标因子
time setting 时间整定
time-settlement curve 时间-沉降曲线
time sharing 分时
time slice 时间片
time switch 定时开关,时间开关
time tagging 时间标记
time-to-amplitude converter (TAC) 时间-幅度变换器
time-to-number converter 时间-数字变换器
time undervoltage protection 时限低电压保护(装置)
time variable 时间变量
time-varying coefficient 时变系数
time-varying field 时变场
time-varying parameter 时变参数
time-varying reactivity 时变反应性

time-varying system 时变系统
time yield limit 持久屈服极限
timing capacitor 计时电容器
timing circuit 定时电路
timing contactor 定时接触器
timing magneto 定时磁电机
timing motor 定时电动机
timing pulse 定时脉冲
timing track 同步磁道
timing voltage 定时电压
timing wave 定时波
tinning 镀锡
tinsel conductor 铜皮线
tip of blade 叶尖
tip clearance 叶顶间隙
tip diameter 顶圆直径
tip-hub ratio 轮毂比
tip jack 尖头插孔
tip speed ratio 叶尖速比
titanate 钛酸盐
titanium dioxide （二）氧化钛
T joint T 型接头
TOC (total organic carbon) 总有机碳(水质参数)
toe crack 焊趾裂纹
Toepler machine 托普勒电机(静电发生器)
toggle circuit 触发电路
toggle flip-flop 计数型触发器
toggle switch 拨动开关
toggle switched capacitor 拨动式开关电容
tomograph 层析 X 射线照相机
tomography 层析 X 射线照相术
tone-burst generator 单音脉冲发生器
tone generator 音频发生器,音频振荡器
tong-test ammeter 钳式电流表,夹线式电流表
tongue and groove joint 榫槽接合
tongue pattern 舌状花样
tooth 齿
tooth coupling 齿式联轴器
toothed armature 有齿电枢
toothed bar 齿条

toothed wave 锯齿波
tooth harmonic 齿谐波
tooth pitch 齿距
tooth ripple 齿(形)波纹
tooth saturation 齿部饱和
tooth sector 扇形齿轮
tooth-tip reactance 齿尖漏抗
top bar 上层线棒;(双鼠笼电机的)上鼠笼条
top chord 上弦
top elevation 顶面标高
top-fired burner 顶置喷燃器
top flange 上法兰
top heater 末级加热器
top-hinged window 上悬窗
topogram 内存储信息位置图示
topographic feature 地貌
topological analysis 拓扑分析
topping governor 主控调速器
topping turbine 前置(式)汽轮机
top slab 顶板
topsoil 表土
top-supported boiler 顶部悬吊支承锅炉
top turbine 前置(式)汽轮机
top yoke 顶轭
torch corona 火焰式电晕
torch igniter 火炬点火器
torch oil gun 点火油枪
toroidal cavity resonator 环形空腔谐振器
toroidal memory core 环形存储磁芯
toroidal transformer 环(状铁)芯变压器
toroidal winding 环形绕组
torque booster 增扭器
torque calculation 转矩计算
torque compensator 转矩补偿器
torque converter transmission 变扭器传动
torque current 转矩电流
torque limitation 转矩限制
torque magnetometer 转矩磁力计
torque motor 力矩电动机
torque pickup 转矩传感器

torquer 力矩器
torque ripple 转矩波动
torque rotary actuator 力矩旋转执行机构
torque shaft 扭转轴
torque-slip characteristic 转矩-滑差特性
Torricellian vacuum 托里拆利真空
torsion 扭力；扭转
torsional critical speed （电机的）扭转临界转速
torsional damper 输电线防震锤；扭力阻尼器
torsional oscillation 扭转振荡
torsional pendulum 扭摆
torsional reinforcement 抗扭配筋
torsion angle 扭转角
torsion meter 扭力计
TOS (tape operating system) （磁）带操作系统
TOT (turbine outlet temperature) 涡轮出口温度
total absorptivity 总吸收率
total amplitude 总振幅，全振幅，总幅值
total capacitance 总电容，全电容
total current 全电流
total differential 全微分
total dissolved solids (TDS) 总含盐量，总溶解固体
total electric load 总电负荷
total exchange capacity 全交换容量
total flow 总流量
total flux 总通量
total harmonic distortion 全谐波畸变
total head 总水头，总压头
total heating surface 总受热面
total heating value 总热值
total impedance 总阻抗
total input impedance 总输入阻抗
total load rejection 全甩负荷
total moisture 总水分，全水分
total organic carbon (TOC) 总有机碳(水质参数)
total radiation fluxmeter 全辐射通量计
total radiation pyrometer 全辐射高温计
total solids 全固形物
total step iteration 整步迭代
tough copper 韧铜
toughened glass 钢化玻璃
tough metal 韧性金属
toughness test 韧性试验
towel rack 毛巾架
tower 杆塔；塔架(输电线)
tower boiler 塔型锅炉
tower reclaiming system 塔式回收系统
tower shadow effect 塔影效应
towing plate 牵引板
Townsend discharge 汤森放电
TR (test run) （测）试运行
trace analysis 痕量分析
tracer atom 示踪原子
tracer element 示踪元素
tracer gas 示踪气体
track 电痕
track-and-hold amplifier 跟踪与同步放大器
track circuit 轨道电路
track density 道密度
track gauge 轨距尺
tracking circuit 跟踪电路
tracking erosion 起痕蚀损
tracking filter 跟踪滤波器
tracking network 跟踪网络
track relay 轨道继电器
track return circuit 轨道回流电路
track scale 轨道衡
track switch 道岔
traction boiler 牵引式锅炉
traction current 牵引电流
traction engine 牵引式发动机
traction generator 牵引(用)发电机
traction load 牵引负载

traction motor 牵引电动机
traffic 运输量;通信量,业务量
traffic distributor 通信业务分配器
trailer block 随附信息组
trailing brush 后倾式电刷
trailing cable 拖曳电缆
trailing edge (TE) 后缘
trailing-edge delay （脉冲）后沿延迟
trailing-edge shock 尾缘激波
trailing section loss 出口边损失
trailing shock wave 尾冲波,尾激波
trailing vortex 尾随涡
trajectory bucket 挑坎
tramegger 高阻表,兆欧表,摇表
transadmittance 跨导纳,互导纳
transceiver （无线电）收发器
transceiver data link 收发器数据链路
transconductance 跨导
transcriber 转录器
transducer 传感器,转换器,换能器
transductor 饱和电抗器;磁放大器
transductor amplifier 磁放大器
transductor-controlled generator 磁放大器控制发生器
transfer admittance 转移导纳
transfer bus 换接母线,切换母线
transfer canal 输送渠道
transfer capacitance 跨接电容
transfer circuit 转移电路
transfer constant 传输常数,转移常数
transfer contact 转换触点,切换触点
transfer control register 传送控制寄存器
transfer curve （饱和电抗器）静特性曲线
transfer factor 传递因数
transfer flask 传送容器
transfer function amplifier 传递函数放大器
transfer function analyzer 传递函数分析器
transfer immittance 传递导抗
transfer impedance 转移阻抗,传递阻抗
transfer machine 连续自动生产线
transfer mechanism 传递机构,传动机构
transfermium element 超锿元素
transfer order 转移指令
transfer pipet 移液管
transfer pump 输运泵
transfer ratio 传递比
transferred tripping 远动跳闸,远方跳闸
transfer relay 切换继电器
transfer strip 切换片
transfer surface 传递面,传质面
transfer switch 转接开关
transfer-type heat exchanger 传递式换热器
transfer valve 切换阀
transfer voltage ratio 转换电压比
transfer winding 成形绕组
transfluxor 多孔磁芯存储器
transformation equation 变换方程
transformation group 变换群
transformation point 转变点,临界点
transformation ratio 变（压）比
transformation-ratio meter 变比计
transformation temperature 相变温度,相变点,临界点
transformer 变压器;变换器
transformer action 变压器作用
transformer bank 变压器组
transformer booster 升压变压器
transformer breaker 变压器断路器
transformer bushing 变压器套管
transformer cabin 变压器室

transformer casing 变压器外壳
transformer coil 变压器线圈
transformer cooler 变压器冷却器
transformer core 变压器铁芯
transformer-coupled amplifier 变压器耦合放大器
transformer-coupled load 变压器耦合负荷
transformer-coupled oscillator 变压器耦合振荡器
transformer-coupled pulse amplifier 变压器耦合脉冲放大器
transformer coupling 变压器耦合
transformer differential protection 变压器差动保护
transformer drop compensator 变压器压降补偿器
transformer effect 变压器效应
transformer electromotive force 变压器电动势
transformer equivalent circuit 变压器等效电路
transformer-fed system 经变压器的供电系统
transformer feeder protection 变压器馈线保护
transformer lead 变压器引线
transformerless multiplex switch 无变压器多接开关
transformerless power supply 无变压器电源,直供电源
transformer magnetic shunt 变压器磁分路
transformer matching 变压器匹配
transformer network 变压器网络
transformer oil 变压器油
transformer oil cooler 变压器油冷却器
transformer output 变压器输出
transformer pillar 变压器塔柱
transformer ratio 变压器比;变流比
transformer-ratio bridge 变压比电桥
transformer reactance 变压器电抗

transformer relay 变压器继电器
transformer rotor (旋转变压器的)变压器转子
transformer sheet 变压器硅钢片
transformer stamping 变压器冲片
transformer station 变电站,变电所
transformer steel 变压器硅钢
transformer substation 变电所,变电站
transformer tank 变压器箱
transformer tap 变压器分接头
transformer turn ratio 变压器匝（数)比
transformer utilization factor 变压器利用系数,变压器利用率
transformer vault 变压器(拱顶)室
transformer voltage 变压器电压
transformer winding 变压器绕组
transformer yoke 变压器磁轭
transforming station 变电所,变电站
transgranular corrosion 穿晶腐蚀
transgranular cracking 穿晶破裂
transgranular fracture 穿晶断裂
transient availability 瞬时可用率,瞬时可用度
transient behaviour 瞬时性态,瞬变特性
transient boiling 过渡(工况)沸腾,瞬态沸腾
transient-cause forced outage 暂态性强迫停运
transient component 瞬态分量
transient condition 瞬态工况
transient creep 暂态蠕变
transient critical flow 瞬态临界流
transient current 瞬态电流
transient current offset 瞬态电流补偿
transient curve 瞬态曲线,过渡过程曲线
transient-decay current 瞬衰电流
transient divider 瞬态分压器

transient drag 瞬时阻力
transient equilibrium 瞬态平衡,暂态平衡
transient fault 瞬时故障,短暂故障
transient flux linkage 瞬间磁链
transient forced outage 瞬时强迫停运
transient generator 暂态发电机
transient oscillation 瞬态振荡
transient overload current 瞬时过载电流
transient overvoltage 瞬态过电压
transient overvoltage counter 瞬态过电压计数器
transient peak 瞬时峰值
transient phenomenon 瞬变现象
transient property 瞬时特性
transient pulse 瞬态脉冲
transient reactance 瞬态电抗
transient recovery voltage 瞬态恢复电压
transient response 瞬态反应,瞬态响应
transient short-circuit time constant 瞬态短路时间常数
transient stability 瞬态稳定(度),暂态稳定(度)
transient state 瞬态,暂态
transient stress 瞬态应力
transient thermal impedance 瞬态热阻抗
transient thermal response 瞬时热响应
transient thermal stress 瞬态热应力
transient vibration 瞬态振动
transient voltage 瞬态电压
transient working area 暂时工作区
transient zone 过渡区
transilog 晶体管逻辑电路
transimpedance 互阻抗
transistor 晶体管
transistor amplifier 晶体管放大器
transistor analysis recording equipment (TARE) 晶体管分析记录设备
transistor and component tester (TACT) 晶体管及元件测试仪
transistor bias circuit 晶体管偏压电路
transistor chopper 晶体管斩波器
transistor circuit 晶体管电路
transistor collector 晶体管集电极
transistor-coupled logic (TCL) 晶体管耦合逻辑
transistor-coupled transistor logic (TCTL) 晶体管耦合晶体管逻辑
transistor current steering logic (TCSL) 晶体管电流导引逻辑
transistor decade counter 晶体管十进制计数器
transistor demodulator 晶体管解调器
transistor-diode logic 晶体管-二极管逻辑
transistor-diode-transistor logic 晶体管-二极管-晶体管逻辑
transistor display and data processing system 晶体管显示器及数据处理系统
transistor electronics 晶体管电子学
transistor equation 晶体管方程
transistor flip-flop 晶体管触发器
transistor logic circuit 晶体管逻辑电路
transistor measuring oscilloscope 晶体管测量示波器
transistor memory 晶体管存储器
transistor motor 晶体管电动机
transistor multivibrator 晶体管多谐振荡器
transistor oscillator 晶体管振荡器
transistor parameter 晶体管参数
transistor relay 晶体管(式)继电器

transistor-resistor logic 晶体管-电阻器逻辑
transistor-resistor-transistor logic 晶体管-电阻器-晶体管逻辑
transistor servo preamplifier 晶体管伺服前置放大器
transistor switch 晶体管开关
transistor torque meter 晶体管转矩仪
transistor trigger 晶体管触发器
transistor voltage stabilizer 晶体管电压稳定器,晶体管稳压器
transistor voltmeter 晶体管伏特计,晶体管电压表
transition 转换,转接;跃迁
transition anode 过渡阳极
transition band 过渡(频)带
transition boiling 过渡沸腾
transition coefficient 过渡系数
transition-coil 过渡线圈,分流线圈
transition condition 过渡状态
transition contact 过渡触头
transition current 过渡电流
transition curve 过渡曲线,转变曲线
transition energy 转变能;跃迁能
transition impedance 过渡阻抗
transition joint 过渡接头
transition region 过渡区
transition resistor 过渡电阻器(变压器抽头转换用)
transition ring (锻件)过渡环
transition state 过渡态
transition temperature 转变温度
transition time 跃迁时间
transit-time tube 速调管,渡越时间管
translawrencium element 超铹元素
translucent body 半透明体
transmission adaptor 传输转接器
transmission and distribution (T & D) 输电与配电,输配电
transmission attenuation 传输衰减
transmission capacity 输电能力
transmission channel 传输通路
transmission characteristic 输送特性
transmission coefficient 传输系数;透射系数
transmission control 传输控制
transmission control element (TCE) 传输控制元件
transmission controller (TC) 传输控制器
transmission control protocol (TCP) 传输控制协议
transmission delay 传输延迟
transmission dynamometer 传动式测力计
transmission end 输电端
transmission equivalent 传输当量
transmission factor 透射因数,透射比;传输系数
transmission frequency 传输频率
transmission grid 输电网
transmission interface converter 传输接口转换器
transmission lag 传输滞后
transmission level 传输电平
transmission line 输电线;传输线
transmission medium 传输媒质,传输媒体
transmission network 输电网
transmission path 传递通路,传输通道,传输路径
transmission performance 传输性能
transmission protocol 传输协议
transmission pulse 传输脉冲
transmission range 传输范围;输电范围
transmission rate 传输速率
transmission ratio 传动比,传输比
transmission regulator 传输调节器
transmission route 输电路线;传输路线,传输路由

transmission semiconductor detector 穿透式半导体探测器
transmission subsystem interface 传输子系统接口
transmission system 输电系统
transmission tower 输电(线路)塔
transmission voltage 输电电压
transmission wire 输电线
transmissivity 透射率
transmittance 透射比,透射因数
transmitting core 发送磁芯
transmitting crystal 发送晶体
transmitting element 发送元件
transmitting rheostat 传送变阻器
transmitting selsyn 自动同步发送器
transmitting transducer 发射换能器
transmitting transformer 输电变压器
transmitter distributor (TD) 发射机分配器
transmityper 导航(光电)信号发送机
transnobelium element 超锘元素
transonic aerodynamics 跨音速空气动力学
transonic cascade 跨音速叶栅
transonic compressor 跨音速压气机
transparency 透明(度)
transpiration-cooled blade 发散冷却叶片
transplutonium 超钚
transport coefficient 运输系数
transport delay unit 运输延迟单元
transport mean free path 输运平均自由程
transport operator 输运算符
transposed conductor 换位导线
transposed integral equation 转置积分方程
transposed matrix 转置矩阵,换位矩阵
transposed winding 换位绕组
transposition 换位;置换
transposition cycle 换位循环
transposition insulator 换位绝缘子,交叉绝缘子
transposition interval 换位间隔,交叉间隔
transposition section 交叉区,交叉段
transposition tower 换位塔,交叉塔
transreactance 互抗(互阻抗的虚数部分)
transrectification 阳极检波
transrectification factor 阳极检波系数
transrectifier (电子管)板极检波器,阳极检波器
transresistance 互阻(互阻抗的实数部分)
transsusceptance 互(导)纳(互导纳的虚数部分)
transtat 自耦变压器,可调变压器
transthorium element 超钍元素
transudate 渗出液
transuranium element 超铀元素
transverse arrangement 横向布置
transverse axis 横轴
transverse buckling 横向屈曲
transverse condenser 横向布置凝汽器
transverse differential protection 横差保护,横联差动保护
transverse diffraction 横向衍射
transverse electromagnetic wave 横电磁波
transverse expansion joint 横向伸缩缝
transverse field 横(向)场
transverse flow 横向流
transverse flux heating 横向磁通加热
transverse focusing 横向聚焦
transverse force 横向力

transverse fracture 横向断裂
transversely-finned fuel element 带横向肋的燃料元件
transverse magnetic field 横向磁场
transverse magnetic wave 横磁波
transverse magnetization 横(向)磁化
transverse mixing 横向混合
transverse movement 横向运动
transverse oscillation 横向振荡
transverse piezoelectric effect 横向压电效应
transverse pitch 横向节距;端面齿距
transverse reinforcement 横向钢筋
transverse resolution 横向分辨力
transverse section 横截面,横断面
transverse slot 横向槽
transverse strain 横向应变
transverse strength 横向强度
transverse stress 横向应力
transverse vibration 横向振动
transverse wedge 横楔
transverter 换流器
trap circuit 陷波电路
trapezoid-shaped slot 梯形槽
trapezoid truss 梯形桁架
trap-water purification 泄放水净化
trass cement 火山灰水泥
travel 行程
travelling cable 电梯用电缆
travelling cathetometer 活动测高计
travelling crane 移动起重机
travelling field 行波场,行移场
travelling grate 链条炉排
travelling hoist 移动式启闭机,移动式卷扬机
travelling overvoltage 行移过电压
travelling phase 前进相位,移动相位

travelling shuttering 移动式模板
travelling soot blower 移动式吹灰器
travelling wave 行波
travelling wave tube (TWT) 行波管
travel switch 行程开关
travel-time curve 时程曲线
traverse point 导线点
triac 双向(三极)晶闸管
trickle resin 滴浸树脂
trifurcating joint 三芯分支接头
trifurcator 三芯分支盒
triggering failure 失触发
trim coil 调整线圈
triode gun (焊接)三极管枪
trochoid (余)摆线
true porosity 全气孔率
trunk cut-off (TCO) 干线切断
T-type antenna T型天线
tube array 管排
tube bank 管束,管排,管组
tube bender 弯管机
tube bundle 管束,管簇
tube cap (电子管)管帽
tube clip 管卡
tube closure 管盖
tube connection 管连接
tube connector 管接头;(导线)连接管
tube-lined boiler wall 敷管炉墙
tubular coil 管形线圈
tubular conductor 管形导线,管形导体
tubular discharge lamp 管形放电灯
tubular heater 管式加热器
tubular plate 管式极板
tubular turbine 贯流式水轮机
tungsten collimator 钨准直器
tunnel diode (TD) 隧道二极管
tunnel furnace 隧道式炉
turbine 涡轮(机)
turbine bypass (TB) 汽轮机旁路
turbine bypass control (TBC) 汽轮机旁路控制

turbine-compressor (TC) 汽轮机-压气机组
turbine cylinder 涡轮汽缸
turbine-driven feed pump 汽轮给水泵
turbine expander 涡轮膨胀机
turbine loss 汽轮机损失
turbine outlet temperature (TOT) 涡轮出口温度
turbine over-speed protection 汽轮机超速保护
turbine vibration protection 汽轮机振动保护
turbo boiler feed pump (TBFP) 汽动锅炉给水泵
turbo-compressor 涡轮压缩机
turbo-generator 汽轮发电机
turbulence 湍流,紊流
turbulent flow 紊流,湍流
turn compensation 匝数补偿
turn insulation 匝绝缘
turntable 转盘
turn-to-turn test 匝间试验
twinax 屏蔽双线馈线
twin-bed (沸腾炉)双床的
twin contact 双触点
twin(-core) cable 双芯电缆
twin-core current transformer 双芯电流互感器
twin crankshaft engine 双曲轴发动机
twin-delta Δ-Δ(接法),双 Δ(连接)
twin diode 双二极管
twin electrode 双电极焊条
twin engine 双联发动机
twin-fluid atomization 双流体雾化
twin furnace 双炉膛
twin insulator strings 双联绝缘子串
twin probe 双针探针
twin-screw extruder 双螺杆挤出机
twin-shell condenser 双壳冷凝器,双壳凝汽器

twin-T network 双 T 形网络
twin triode 双三极管
twin-turbine 双流(型)汽轮机
twin-turbine generator 双汽轮发电机
twin twisted strand 双绞股线
twin wire 双股线
twisted blade 扭叶片,弯扭叶片
twisted cord 绞合软线
twisted joint 绞接头
twisted-lead transposition 绞线换位
twisted line 绞线
twisted pair 双绞线
twisting moment 扭(转力)矩
twisting stress 扭应力
twist joint 绞合接头;绞接
twistor 扭量;磁扭线
two-armature motor 双电枢电机
two-bank engine 双列汽缸发动机
two-bearing machine 双轴承型电机
two-bed demineralizer 双床除盐装置
two-circuit receiver 双调谐电路接收机
two-circuit winding 双路绕组
two-coil instrument 双线圈仪表
two-coil relay 双线圈继电器
two-column core 双柱型铁芯
two-component heat pipe 双组分热管
two-core excitation winding 双铁芯励磁绕组
two-core reactor 双芯反应堆
two-core switch 双磁芯开关
two-core transformer 双芯变压器
two-cycle engine 双冲程发动机
two-cycle generator 双周波发电机
two-degree-of-freedom gyroscope 二自由度陀螺仪
two degrees of freedom (TDF) 二自由度
two-film theory 双膜理论
two-flow core 双流程堆芯

two-fluid reactor 双流体反应堆
two-fluid system 双流体系统
two-furnace boiler 双炉体锅炉
two-gang variable capacitor 双联可变电容器
two-group critical equation 双群临界方程
two-group critical mass 双群临界质量
two-group diffusion theory 双群扩散理论
two-group model 双群模型
two-group perturbation theory 双群微扰理论
two-group theory 双群理论
two-gun oscillograph 双枪示波器
two-input adder 双输入加法器,半加(法)器
two-input gate 双输入门
two-input servo 双输入伺服
two-input subtracter 双输入减法器,半减(法)器
two-input switch 双输入开关
two-layer winding 双层绕组
two-legged transformer 双柱式变压器
two-loop servomechanism 双回路伺服机构
two-neck(ed) distilling flask 双颈蒸馏瓶
two-pass boiler 双回程锅炉
two-pass condenser 双流程凝汽器
two-pass cooling 双路冷却
two-pass core 双流程堆芯
two-phase boundary layer 两相边界层
two-phase circuit 两相电路
two-phase flow 两相流,二相流
two-phase flow regime 两相流动工况
two-phase fluid 两相流体
two-phase four-wire system 两相四线制
two-phase generator 两相发电机
two-phase ground fault 两相接地故障
two-phase mixing 两相交混
two-phase servomotor 两相伺服电机
two-phase state 两相态
two-phase system 两相系统
two-phase theory 二相理论
two-phase transformer 两相变压器
two-piece wedge 分半槽楔
two-point extraction cycle 两段抽汽循环
two-pole machine 两极电机
two-pole switch 双极开关
two-port network 二端口网络,二端对网络
two-position action 双位置动作
two-quadrant converter 双象限变流器
two-terminal circuit 二端电路
two-winding transformer 双绕组变压器
TWT (travelling wave tube) 行波管
twystron 行波速调管
tympanum (电话机的)振动膜
type efficiency 型效率
typical cell 典型栅元

UA (uniform array) 一致阵列；一致数组
UACTE (universal automatic control and test equipment) 通用自动控制及测试设备
UADPS (uniform automatic data processing system) 一致自动数据处理系统
UADS (user attribute data set) 用户属性数据集
UART (universal asynchronous receiver/transmitter) 通用异步接收发送设备[器]
UBC (universal buffer controller) 通用缓冲控制器
U-bend specimen U 型弯曲(应力腐蚀)试样
U-bolt U 型螺栓
UCC² (utility control center) 应用管理中心
U-clamp U 型夹
U core U 型磁铁芯
UCP (utility control program) 实用控制程序
UCS¹ (universal character set) 通用字符集
UCS² (universal classification system) 通用分类系统
UD¹ (universal dipole) 通用偶极子
UD² (unplanned derating) 非计划降额
UDC (universal decimal classification) 通用十进制分类法
UDF (unit derating factor) 机组降额因数

UDG (unit derating generation) 机组降额发电量
UDH (unplanned derating hours) 非计划降额小时
UERS (unusual event recording system) 异常事件记录系统
UF (unavailable factor) 不可用系数
U-flame furnace U 型火焰炉
UH (unavailable hours) 不可用小时
UHF (ultra-high frequency) 超高频,特高频
UHF ceramic capacitor 超高频陶瓷电容器
UHF correlator 特高频相关器
UHT (ultra-high temperature) 超高温
UHV¹ (ultra-high vacuum) 超高真空
UHV² (ultra-high voltage) 特高(电)压
UHV transmission 特高压输电
UHV transmission line 特高压输电线路
UIOD (user I/O device) 用户输入输出设备
ULB (universal logic block) 通用逻辑块
ULF (ultra-low frequency) 超低频,特低频
ullage 空距；漏损(量),损耗(量)
ultimate analysis 元素分析
ultimate bearing capacity 极限承载力
ultimate bending moment 极限弯矩

ultimate capacity 最大功率；最大容量，极限容量
ultimate composition 元素成分
ultimate compressive strength 极限抗压强度
ultimate creep 蠕变极限
ultimate damping 极限阻尼，极限衰减
ultimate deformation 极限变形
ultimate disposal 最终处置，最终处理
ultimate elongation 极限伸长
ultimate filter 终端过滤器
ultimate fuel burn-up 最终燃料消耗
ultimate gain 最大增益
ultimate heat sink 最终热阱
ultimate heat sink basin 最终热阱水坑
ultimate installed capacity 最大装机容量
ultimate limit state 承载能力极限状态
ultimate load 极限负荷，极限载荷，最大负荷
ultimate load design 极限载荷设计
ultimately controlled variable 最终被控变量，最终受控变量
ultimate maximum 极限最大值
ultimate mechanical strength 极限机械强度
ultimate output 最大出力，最大输出
ultimate pH 极限 pH
ultimate pressure 最终压力
ultimate regenerative cycle of saturated steam 饱和蒸汽极限回热循环
ultimate resistance 极限阻力
ultimate sampling unit 最后取样单位
ultimate sensibility 最高灵敏度
ultimate sensitivity 最高灵敏度
ultimate storage 最终贮存
ultimate storage drum 最后贮存桶
ultimate strain 极限应变
ultimate strength 强度极限，极限强度
ultimate strength design 极限强度设计
ultimate temperature 最终温度
ultimate tensile strength 极限拉伸强度，极限抗拉强度
ultimate tensile stress 极限(抗)拉应力
ultimate tension 极限张力
ultimate vacuum 极限真空
ultimate value 极限值
ultimate waste disposal 废物最终处置
ultimate working capacity 极限工作容量
ultimate yield 最终产量
ultra-audible frequency 超声频率，超音频
ultracentrifuge 超速离心机
ultra-clean air system 超净空气系统
ultracold neutron 超冷中子
ultrafast neutron 超快中子
ultrafilter 超滤器
ultrafiltrate 超滤液
ultrafiltration 超(过)滤
ultrafiltration membrane 超滤膜
ultrafiltration membrane process 超滤膜法
ultrafiltration process 超滤法
ultrafiltration system 超滤系统
ultrafine dust 特细粉尘
ultrafine enamelled wire 特细漆包线
ultrafine glass-coated wire 特细玻璃丝包线
ultrafine ion-exchange resin 超微离子交换树脂
ultrafine particle 超微颗粒，特细粒子
ultrafine soot 特细煤烟
ultraharmonic 超谐波

ultra-high frequency (UHF) 超高频,特高频(300—3 000 兆赫)
ultra-high-frequency tuner 超高频调谐器
ultra-high purity 超高纯度
ultra-high-purity helium atmosphere 超高纯度氦气氛(层)
ultra-high resolution 超高分辨力
ultra-high-speed transient wave form analyzer 超高速瞬时波形分析器
ultra-high temperature (UHT) 超高温
ultra-high-temperature reactor 超高温反应堆
ultra-high vacuum (UHV) 超高真空
ultra-high vacuum valve 超真空阀
ultra-high voltage (UHV) 特高(电)压
ultra-low emission 超低污染排放
ultra-low frequency (ULF) 超低频,特低频
ultra-low-temperature thermistor 超低温热敏电阻(器)
ultra-microanalysis 超微(量)分析
ultramicrobalance 超微量天平
ultramicrochemistry 超微量化学
ultramicro-determination 超微量测定
ultramicrometer 超测微计
ultramicroscope 超显微镜
ultramicroscopic dust 特细微尘
ultrapure water 高纯水,超纯水
ultra-purity 超纯度
ultrared ray 红外线
ultra-sensitive leak testing method 超灵敏检漏法
ultrashort wave 超短波
ultrasonator 超声振荡器
ultrasonic alarm system 超声波报警系统
ultrasonic atomizer 超声波雾化器
ultrasonic beam 超声束
ultrasonics 超声学
unavailable factor (UF) 不可用系数
unavailable hours (UH) 不可用小时
unconnected network 非连通网络
under-current relay 欠电流继电器
underflow 底流,下漏
underground cable 地下电缆
underload 欠载
under-reaching pilot protection 欠范围式纵联保护
under-voltage protection 欠电压保护
unearthed voltage transformer 不接地电压互感器
uniform array (UA) 一致阵列;一致数组
uniform automatic data processing system (UADPS) 一致自动数据处理系统
uninterrupted power supply (UPS) 不间断电源
unit connection diagram 单元接线图
unit derating factor (UDF) 机组降额因数
unit derating generation (UDG) 机组降额发电量
universal asynchronous receiver/transmitter (UART) 通用异步接收发送设备[器]
universal automatic control and test equipment (UACTE) 通用自动控制及测试设备
universal buffer controller (UBC) 通用缓冲控制器
universal character set (UCS) 通用字符集
universal classification system (UCS) 通用分类系统
universal decimal classification (UDC) 通用十进制分类法

universal dipole (UD) 通用偶极子

universal instrument 万用表

universal logic block (ULB) 通用逻辑块

universal motor 交直流两用电动机

unlined tunnel 无衬砌隧洞

unplanned derating (UD) 非计划降额

unplanned derating hours (UDH) 非计划降额小时

unsymmetrical current 不平衡电流

unusual event recording system (UERS) 异常事件记录系统

upper reservoir 上池,上水库

UPS (uninterrupted power supply) 不间断电源

urban power network planning 城市电网规划

useful heat 有效热

user attribute data set (UADS) 用户属性数据集

user I/O device (UIOD) 用户输入输出设备

utility control center (UCC) 应用管理中心

utility control program (UCP) 实用控制程序

utilized flux 利用光通量

U-tube U型管

U-tube manometer U型管压力计

VA (volt-ampere) 伏安
vacancy chromatography 空穴色谱法
vacuum 真空
vacuum accumulator 真空贮气筒
vacuum correction 真空修正
vacuum drying oven 真空干燥箱
vacuum evaporation 真空蒸发
vacuum gauge 真空表,真空规,真空压力计
vacuum melting 真空熔炼,真空熔融
vacuum sensor 真空传感器
vacuum system 真空系统
vacuum transducer 真空传感器
vacuum valve 真空阀
value of scale division 标尺分格值
value of scale mark 标度值
valve 阀;热离子管
valve base 阀基
valve blocking 阀闭锁
valve body 阀体
valve capacity 阀容量
valve element 阀片
valve outlet 阀出口
valve plug 阀芯
valve reactor 阀电抗器
valve seat 阀座
valve shaft 阀轴
valve sizing 阀尺寸计算,阀口径计算
valve stem 阀杆
valve stem unbalance 阀杆不平衡
vane 叶片;风向标
vane-type circuit 叶片型电路
vane-type relay 扇型继电器
vaporization 汽化
varactor 变容二极管
var-hour meter 无功电度表,乏时计
variable 变量
variable-angle internal reflection element 可变角内反射元件
variable-angle probe 可变角探头
variable-area flowmeter 变截面流量计,变面积流量计
variable-area method 变面积法
variable capacitance 可变电容
variable capacitance diode 变容二极管
variable capacitor 可变电容器
variable inductor 可变电感器
variable gain 可变增益,可变放大系数
variable gain method 可变增益法
variable-head flowmeter 变压头流量计
variable parameter operation 变工况运行
variable pressure 变压,变化压力
variable reluctance transducer 变磁阻式传感器
variable speed rotor 变转速风轮
variable standard capacitor 可变标准电容器
variable structure control system 变结构控制系统
variac 自耦变压器
variator 变化器
varicap 变容二极管
variocoupler 可变耦合器

variometer 变感器,可变电感器;磁变仪
varistor 压敏变阻器
varmeter 无功功率计,乏表
varnish 清漆
varnished fabric 浸漆织物
VCCO (voltage-controlled crystal oscillator) 压控晶体振荡器
VCXO (voltage-controlled X-tal oscillator) 压控晶体振荡器
VDA (video distribution amplifier) 视频分布放大器
VDR (voltage-dependent resistor) 压敏电阻器
velocity 速度
velocity-area method 速度面积法
velocity distribution 速度分布
velocity error 速度误差
velocity error coefficient 速度误差系数
velocity feedback 速度反馈
velocity focusing 速度聚焦
velocity limiter 速度限制器
velocity measuring thermistor 测速热敏电阻器
velocity of approach factor 渐近速度系数
velocity pickup 速度传感器
velocity sensor 速度传感器
velocity transducer 速度传感器
velocity-type water meter 速度式水表
velodyne 调速发电机
vena contracta 缩脉,流颈
vent hole 排气孔
ventilated psychrometer 通风干湿表
ventilated thermometer 通风温度表
ventilation rate 换气率,通风率
ventilator 通风器
Venturi nozzle 文丘里喷嘴
Venturi tube 文丘里管
vernier 游标
vertical angle 垂直角
vertical axis 竖轴,垂直轴
vertical balancing machine 立式平衡机
vertical current flow method 直角通电法
vertical limit 垂直极限
vertical scanning 垂直扫描
vertical tail 垂直尾翼
vertical temperature profile radiometer (VTPR) 温度垂直廓线辐射仪
vertical visibility 垂直能见度
VHF omnidirectional range (VOR) 伏尔,甚高频全向信标
vibrating grate 振动炉排
vibrating reed instrument 振簧系仪表
vibrating sample magnetometer (VSM) 振动样品磁强计
vibrating viscometer 振动黏度计
vibrating wire drawing force meter 振弦式拉力计
vibrating wire force transducer 振弦式力传感器
vibrating wire tensiometer 振弦式张力计
vibrating wire tension transducer 振弦式张力传感器
vibrating wire torque measuring instrument 振弦式转矩测量仪
vibrating wire torque transducer 振弦式转矩传感器
vibrating wire transducer 振弦式传感器
vibration 振动
vibration amplitude 振幅
vibration analyzer 振动分析仪
vibration controller 振动控制仪
vibration cylinder barometer 振筒式气压表
vibration cylinder pressure transducer 振筒式气压传感器
vibration error 振动误差
vibration exciter 激振器
vibration galvanometer 振动检流计

vibration generator 激振器，振动发生器
vibration isolation 隔振
vibration isolator 隔振器
vibration measurement by absolute method 绝对测振法
vibration measurement by mechanical method 机械测振法
vibration measurement by relative method 相对测振法
vibration monitor 振动监视器，振动监测仪
vibration pickup 拾振器
vibration sensor 振动传感器
vibration severity 振动烈度
vibration table 振动台
vibration test 振动试验
vibration transducer 振动传感器
vibrator 振动器，激振器
vibrometer 振动计
Vickers hardness number 维氏硬度值
Vickers hardness penetrator 维氏硬度压头
Vickers hardness tester 维氏硬度计
video distribution amplifier (VDA) 视频分布放大器
virtual image 虚像
virtual image mass spectrometer 虚像质谱计
virtual memory 虚拟存储(器)
viscometer 黏度计
viscosity 黏度
viscosity balance 黏度天平
viscosity sensor 黏度传感器
viscosity transducer 黏度传感器
visibility 能见度，可见度
visibility marker 能见度目标物
visibility meter 能见度表
visibility object 能见度目标物
visible function 视见函数
visible light sensor 可见光传感器
visible light transducer 可见光传感器
visible radiation 可见光辐射
visible spectral remote sensing 可见光遥感
visual axis 视轴
visual field 视场
visual inspection 外观检验
visual range 视程，能见范围
visual range meter 视距测定仪
VLF electromagnetic receiver 甚低频电磁仪
voltage amplifier 电压放大器
voltage centre 电压中心
voltage circuit 电压回路
voltage constant 电压常数
voltage-controlled crystal oscillator (VCCO) 压控晶体振荡器
voltage-controlled gate module 电压控制门组件
voltage-controlled X-tal oscillator (VCXO) 压控晶体振荡器
voltage-dependent resistor (VDR) 压敏电阻器
voltage divider 分压器
voltage divider with auxiliary branch circuit 具有辅助支路的分压器
voltage drop 电压降
voltage error 电压误差
voltage fluctuation 电压波动
voltage matching transformer 电压匹配互感器
voltage ratio 电压比，分压比
voltage ratio of a capacitor divider 电容器式分压器的电压比
voltage regulation 电压调整(率)
voltage sensitivity 电压敏感性，电压灵敏度
voltage sensor 电压传感器，电压敏感器
voltage stabilizing circuit 稳压电路
voltage transducer 电压传感器
voltage transformer 电压互感器
voltammetry 伏安法

volt-ampere (VA) 伏安
volt-ampere-hour meter 视在功率电度表,伏安时计
volt-ampere meter 伏安表,视在功率表
voltmeter 电压表,伏特表
volume conductive humidity sensor 体电导式湿度传感器
volume conductive humidity transducer 体电导式湿度传感器
volume flow 体积流量总量
volume flow rate 体积流量
volume thermodilatometry 体膨胀法
volumetric chromatography 体积色谱法
volumetric flowmeter 容积式流量计
volumetric method 容积法
volumetric water meter 容积式水表
volume viscosity 体积黏度
VOR (VHF omnidirectional range) 伏尔,甚高频全向信标
vortex flowmeter 旋涡流量计
vortex precession flowmeter 旋进流量计
vortex-shedding flowmeter 涡街流量计
vortex-shedding flow transducer 涡街流量传感器
vortex street 涡街
VSM (vibrating sample magnetometer) 振动样品磁强计
VTPR (vertical temperature profile radiometer) 温度垂直廓线辐射仪

wall superheater 墙式过热器
Ward-Leonard system 沃德-伦纳德系统
warm water discharge 温排水
warning alarm 警报
warning circuit 报警电路
warning device 报警装置
warning light 报警(信号)灯
warning limit 警戒界限
warning signal 警告信号
warp 翘曲
waste acid 废酸
waste activated sludge 废活性污泥
waste air 废气
waste alkali 废碱
waste back-cycling 废液返回循环
waste bank 废物堆
waste boiler 废物锅炉
waste burial 废物埋葬
waste burial ground 废物埋葬场
waste calcining facility 废物焚烧装置
waste canal 退水渠
waste carbon 废煤
waste category 废物种类
waste chemical reagent 废化学试剂
waste chrome liquor 废铬液
waste coal 废碳
waste cock 泄放旋塞
waste collector system 废物收集系统
waste condensate pump 废凝结水水泵
waste conditioning 废物形态调整
waste confinement (放射性)废物封隔
waste container 废物容器
waste containment (放射性)废物封存
waste control 废物控制
waste convertion technique 废物转换技术
waste crude oil 废原油
waste cyanide 废氰化物
waste decay tank 废物衰变箱
waste demineralizer 废液除盐器
waste demineralizing plant 废液除盐装置
waste discharge 排废
waste disposal 废物处置
waste disposal basin 废水池
waste disposal beneath the ocean floor 废物海床下处置
waste disposal by nuclear transmutation 废物核嬗变处置
wasted work 耗功
waste evaporative reflux condenser 废液蒸发回流冷凝器
waste fluid 废液
waste fluid burning plant 废液燃烧装置
waste form 废物形态
waste fuel 废燃料
waste gas analysis 废气分析
waste gas burning 废气燃烧
waste gas cleaning 废气净化
waste gas cleaning plant 废气净化设备
waste gas cleaning system 废气净化系统
waste gas compressor 废气压缩机

waste gas desulphurization 废气脱硫
waste gas disposal system 废气处置系统
waste gas emission standard 废气排放标准
waste gas of coal pulverization 磨煤废气
waste gas pollution control 废气污染防治
waste gas supercharger 废气增压器
waste gas superheater 废气过热器
waste gas system 废气系统
waste gas treatment 废气处理
waste gas vapour trap 废气蒸气捕集器
waste heat flue 废热烟道
waste heat loss 废热损失
waste heat recovery 废热回收，余热回收
waste heat removal system 废热排出系统
waste heat superheater 废热过热器
waste heat utilization 废热利用
waste immobilization plant 废物固化工厂
waste liquor recovery 废液回收
waste load 废物负载
waste lye 废碱液
waste management 废物管理
waste material 废物，废料
waste monitoring 废物监测
waste oil 废油
waste oil clot 废油凝块
waste oil disposal 废油处置
waste oxidation basin 废水氧化池
waste package 废物包装
waste pipe 污水管
waste placement 废物安置
waste plant 废物处理设施
waste recovery power plant 废物回收发电厂
waste repository 废物处置库
waste retention system 废物贮存系统
waste retrieval system 废物收回系统
waste sampling cabinet 废物取样箱
waste sand and gravel 废砂石
waste shredder 废物粉碎机
waste silk 废丝
waste sludge 废污泥
waste solidification 废物固化（处理）
waste stabilization pond 废水稳定塘
waste stock 废料
waste storage 废物贮存
waste-storage area 废物贮存区
waste-storage farm 废物贮存场
waste treatment 废物处理
waste treatment equipment 废物处理设备
waste treatment station 废物处理站
waste treatment technique 废物处理技术
waste uranium 废铀
waste valve 排污阀
wastewater collection 废水收集
wastewater collection system 废水收集系统
wastewater composition 废水组成
wastewater dilution 废水稀释
wastewater disinfectant 废水消毒剂
wastewater disposal facility 废水处置设施
wastewater engineering 排水工程
wastewater facility 废水处理设施
wastewater flow 废水流量
wastewater neutralization tank 废水中和箱
wastewater neutralizer 废水中和器

wastewater pipe 废水管
wastewater processing 废水处理
wastewater purification 废水净化
wastewater purification plant 废水净化厂
wastewater rate 废水(排放)率
wastewater reclamation 废水回收
wastewater reclamation plant 废水回收厂
wastewater renovation 中水回用
wastewater reuse 废水复用,废水再用
wastewater sludge 废水污泥
wastewater solid 废水固渣
wastewater sulphide 硫化物废水
wastewater treatment 废水处理
wastewater treatment plant 废水处理厂
wastewater treatment process 废水处理过程
wastewater treatment system 废水处理系统
water absorption 吸水性
water absorptivity 吸水率
water activity meter 水活度测量仪
water admixing installation 混水装置
water-air-cooled machine 间接水冷式空冷电机
water-air ratio 水气比
water analysis 水分析
water aspirator 水泵
water as refrigerant 冷剂水
water atomizer 水雾化器
water baffle 挡水板
water-bath evaporator 水槽蒸发器
water battery 水电池
water blocking 水锁
water body pollution 水体污染
water boiler neutron source (WBNS) 水锅炉中子源
water-borne waste 水载废物
water brake 水力制动器

water budget 水平衡
water carriage 水运
water-carriage system 水运系统
water carry-over (蒸汽中的)夹带水分
water catchment area 集水区
water-cement ratio (w/c) 水灰比
water channel 水道
water chlorination 水加氯处理
water circuit-breaker 断水器
water circulating pump 循环水泵
water circulation 水循环
water circulation coefficient 水循环系数
water circulation detector 水循环检测器
water circulation pipe 水循环管
water circulation pump 水循环泵
water circulation system 水循环系统
water clarification 水澄清作用
water column (WC) 水柱
water-column arrester 水柱避雷器
water-column correction 水柱(高度)校正
water-column pressure 水柱压力
water conditioning 水处理
water contamination 水污染
water content of steam 蒸汽含水量
water control device 水控装置
water control valve 水控制阀
water-cooled condenser 水冷式冷凝器
water-cooled electromagnet 水冷电磁铁
water-cooled forced-oil transformer 水冷强制油循环式变压器
water-cooled furnace 水冷炉膛
water-cooled generator 水冷发电机
water-cooled grate 水冷炉排
water-cooled lattice 水冷栅格
water-cooled machine 水冷电机

water-cooled motor 水冷电动机
water-cooled oil-immersed transformer 水冷油浸变压器
water-cooled oil-insulated transformer 水冷油绝缘变压器
water-cooled reactor 水冷电抗器;水冷反应堆
water-cooled resistor 水冷电阻器
water-cooled rotor 水冷转子
water-cooled transformer 水冷变压器
water-cooled turbine 水冷汽轮机
water-cooled turbogenerator 水冷汽轮发电机
water-cooled wall 水冷壁
water cooler 水冷却器
water-cooling system 水冷(却)系统
water-cooling tower 水冷却塔
water corrosion 水腐蚀
water curing 水养护
water curtain 水幕
water-cycling system 水循环系统
water delivery 送水,供水
water discharge 排水量
water-discharge tube 排水管
water displacer rod (WDR) 挤水棒
water divider 配水器
water ejector 水喷射器,射水抽水器
water engine 水力发动机
water equivalent 水当量
water erosion 水蚀
water-extracting medium 水萃取介质
water feeder 给水器
water-filled fracture 充水裂缝
water film 水膜
water film coefficient 水膜系数
water-film scrubber 水膜除尘器
water filtration 滤水
water flow 水流
water-flow alarm system 水流量警报系统
water-flow circuit 水循环回路
water-flow regulator 水流量调节器
water-flow relay 水流继电器
water flux 水通量
water gap between fuel assemblies 燃料组件间水隙
water gas 水煤气
water gas shift 水煤气转化
water-gauge glass 水表面玻璃
water hammer 水击;水锤
water hammer of heat-supply system 供热系统水击
water hardening 水冷淬火
water hardness 水硬度
water head 水头
water heating 水加热
water heating bath 水加热槽
water immersion test 水浸试验
water immersion wire 潜水电线
water ingress 水浸入(废物库或堆芯)
water injection 注水
water injector 水射器
water inlet 进水口
water intake 进水口
water jacket 水套
water-jacketed condenser 水套冷凝器
water-jet air ejector 射水抽气器
water-jet arrester 水柱避雷器
water-jet condenser 喷水冷凝器
water-jet cutting 水射流切割
water-jet fan 水力引射器
water-jet nozzle 水喷嘴
water-jet pump 水力喷射泵,喷水泵
water lance 喷水枪
water leakage in condenser 冷凝器漏水
water leg (小型锅炉下部的)水夹套
water level 水位
water-level alarm 水位报警
water-level float 水位指示浮子
water-level instrumentation 水位监测仪表

water-level reactivity coefficient 水位反应性系数
water-level recorder 水位记录器
waterline corrosion 水线腐蚀
water load 水负载
water-load power meter 水负载功率计
water logging effect 进水效应(由回路冷却水渗入到元件包壳内部造成)
water-lubricated bearing 水润滑轴承
water management 水务管理
water meter 水表
water-mixing direct connection 混水连接
water-moderated reactor 水慢化反应堆
water-moderated water-cooled reactor 水慢化水冷反应堆
water moderator 水慢化剂
water monitoring 水(质)监测
water motor 水(力发)动机
water of crystallization 结晶水
water outlet 出水口
water penetration 透水(性)
water phantom 水假体
water pipe 水管
water pipe grounding system 水管接地制
water pollution 水(质)污染
water pollution index 水污染指数
water pollution monitor 水污染监测仪
water pressure 水压
water pressure relay 水压继电器
water pretreatment 水预处理
water processing 水处理
waterproof 防水的
waterproof carpet 防水毡
waterproof case 防水外壳
waterproof course 防水层
waterproof electrical equipment 防水型电气设备
waterproof induction motor 防水型感应电动机
waterproofing agent 防水剂
waterproofing powder 防水粉
waterproof instrument 防水型仪表
waterproof insulation 防水绝缘
waterproof layer 防水层
waterproof machine 防水型电机
waterproof mortar 防水砂浆
waterproof motor 防水型电动机
waterproof structure 防水结构
water pump 水泵
water pump house 水泵房
water purification 水净化
water purification plant 净水厂
water purification structure 净水构筑物
water purifying tank 净水槽
water quality analysis 水质分析
water quality control 水质控制
water quality criterion 水质判据
water quality evaluation 水质评价
water quality goal 水质目标
water quality management 水质管理
water quality monitoring 水质监测
water quality pollutant 水质污染物
water quality pollution 水质污染
water quality standard 水质标准
water radioactivity meter 水放射性(活度)测量计
water radiolysis 水辐解
water recirculation 水再循环
water recovery 水回收
water recovery apparatus 水回收设备
water-reducing admixture 减水外加剂
water-reflected system 水反射系统
water removal 脱水
water-resistance load tank 水阻负载箱

water-resistant solution 防水浆
water resistor 水(电)阻器
water return pipe 回水管
water return valve 回水阀
water reuse 水回用
water ring pump 水环泵
water scrubber 水洗涤器
water seal 水封
water seal chamber 水封室
water-sealed gland 水封套
water seal gland 水封(以水为密封介质,防止汽轮机轴端泄漏)
water seal tank 水封箱
water separator 汽水分离装置
waterside corrosion 水侧腐蚀
water sluicing 水力除灰
water-sluicing system 水力除灰系统
water softening (硬)水软化
water softening agent (硬)水软化剂
water softening plant 软水处理装置
water solubility 水溶性
water-soluble halogen 水溶性卤素
water-soluble inhibitor 水溶性缓蚀剂
water-soluble resin 水溶性树脂
water spouted bed 喷水床
water spray cooling 喷水冷却
water spray fire extinguisher 水喷雾式灭火器
water spraying tray 淋水盘
water spray valve 喷水阀
water-submerged motor 潜水型电动机
water supply 供水
water supply engineering 给水工程
water supply pipe 供水管
water supply pump 给水泵
water supply system 给水系统
water surplus 水分盈余
water surveillance network 水监测网

water system 水系统
water tank for heat insulation 隔热水箱
water tank for thermal insulation 隔热水箱
watertight closure 不透水外壳,水密罩
watertight construction 防水结构
watertightness 不透水性
watertight seal 不透水密封
watertight test 水密试验
water-to-steam ratio 水汽比
water toughening 水韧处理
water-to-uranium atomic ratio 水铀原子比
water-to-uranium mass ratio 水铀质量比
water-to-uranium volume ratio 水铀体积比
water-to-water heat exchanger 水-水换热器
water treatment 水处理
water treatment building 水处理厂房
water treatment chemical 水处理化学制品
water treatment equipment 水处理设备
water treatment facility 水处理设施
water treatment plant 水处理厂
water treatment room 水处理室
water treatment system 水处理系统
water tube 水管
water tube boiler 水管(式)锅炉
water turbine generator 水轮发电机
water turbogenerator 水轮发电机
water/uranium volume ratio 水铀体积比
water vapour 水蒸气,水汽
water vapour absorption 水蒸气吸收

water vapour arc welding 水蒸气电弧焊
water vapour content (沸腾炉)水汽含量
water vapour permeability 水蒸气渗透性
water volume 水容量
water wall 水冷壁
water wall circuit 水冷壁回路
water wall cooling 水冷壁冷却
water-walled furnace 水冷壁炉膛
water wall surface 水冷壁面
water wall tube 水冷壁管
water washing 水洗,水力冲洗
water-water heat exchanger 水-水换热器
water-water reactor 水-水反应堆
water-water type heat exchanger 水-水式换热器
water-wave analogy 水波模拟
water-wheel generator 水轮发电机
waterworks 自来水厂
watt 瓦(特)(功率单位)
wattage 瓦数
wattage dissipation 瓦数损耗
wattage output 输出瓦数
wattage rating 额定瓦数
wattage transformer 小功率变压器
watt component 有功部分,有功分量
watt consumption 功率消耗
wattful component 有功部分,有功分量
wattful current 有功电流,有效电流
wattful power 有功功率
watt-hour 瓦(特小)时
watt-hour capacity 瓦时容量
watt-hour constant 瓦时常数
watt-hour demand meter 瓦时需量计
watt-hour efficiency 瓦时效率
watt-hour meter 瓦时计,电度表
watt-in 输入功率

wattless component 电抗部分,虚部
wattless component meter 无功分量表
wattless component watt-hour meter 无功电度表
wattless current 无功电流
wattless power meter 无功功率计
watt loss 功率损耗
wattmeter 瓦特计,功率表
wattmetric relay 瓦特计式继电器
watt-out 输出功率
watt-second 瓦秒
Wauchope type starter 沃克普式起动器,不断路星形-三角形起动器
wave absorber 消波装置,电波吸收体
wave acoustics 波动声学
wave analyzer 谐波分析仪,波形分析器
wave angle 电波仰角,波程角
wave antenna 行波天线
wave attenuation 波衰减
wave changing switch 波长转换开关,波段转换开关
wave channel 波道
wave coil 波绕线圈
wave constant 波长常数
wave converter 波形变换器
wave crest 波峰,波顶
wave detector 检波器
wave differential equation 波动微分方程
wave director 导波体,波导
wave drag 波阻(力)
wave energy 波能
wave energy generator 波能发电机
wave filter 滤波器
waveform 波形
waveform converter 波形变换器
waveform distortion 波形畸变,波形失真

waveform pulse 波形脉冲
waveform recorder 波形记录器
waveform separation 波形分离,波形区分
waveform synthesizer 波形合成器
waveform test 波形试验
wave frequency 波动频率,波频
wavefront 波前,波阵面
wavefront angle 波阵面倾角,波前倾角
wavefront matching 波前匹配,波阵面匹配
wavefront steepness 前陡度
wave function 波函数
wave generator 波发生器
wave group 波群
waveguide 波导(管)
waveguide branching filter 波导分支滤波器
waveguide communication 波导通信
waveguide filter 波导滤波器
waveguide impedance 波导阻抗
waveguide junction 波导结
waveguide klystron 波导型调速管
waveguide magnetron 波导磁控管
waveguide modulator 波导调制器
waveguide reflector 波导反射器
waveguide storage 波导存储器
waveguide switch 波导开关
wave heater (电磁)波加热器
wave impedance 波阻抗
wave interference 波干扰,波干涉
wave launcher 电波发射器,射波器
wavelength 波长
wavelet 小波
wave line 波线
wave loop 波腹
wavemeter 波长计,波频计
wave mode 波模

wave motion 波动,波状运动
wave motor 波力电机
wave node 波节
wave normal 波法线
wave number 波数
wave of translation 平移波
wave packet 波包
wave pattern 波型
wave-power device 波浪发电装置
wave-power generation 波力发电,波浪发电
wave-power station 波力电站,波浪电站
wave process 波动过程
wave propagation 波传播
waver 波段开关;波形转换器
wave radiation (电)波辐射
wave range 波段
wave-range switch 波段开关
wave reflection 波浪反射
wave ridge 波峰
wave selector 波型选择器;波段开关
wave separator 分波器
waveshape 波形
waveshaper 波形形成器
waveshaping 波形形成
waveshaping circuit 整形电路
wave spectrum 波谱
wave staff 测波杆
wave tail 波尾
wave telemetry 电波遥测(术)
wave theory 波动说
wave tilt 波前倾斜
wave train 波列
wave-train frequency 波列频率
wave trap 陷波电路
wave trough 波谷
wave-type vibration 波型振动
wave vector 波矢量
wave velocity 波速
wave voltage 波电压
wave winding 波绕组
wavy flow 波状流
waxed cotton-covered wire 涂蜡纱包线

wax seal 蜡封
WB (write buffer) 写缓冲器
Wb (weber) 韦伯(磁通量单位)
WBC (whole-body counter) 全身计数器
WBD (wide-band data) 宽(频)带数据
WBDL (wide-band data link) 宽(频)带数据链路
WBNS (water boiler neutron source) 水锅炉中子源
W-bridge W 型电桥
WBS (work breakdown structure) 作业分解结构
WC (water column) 水柱
w/c (water-cement ratio) 水灰比
WCA (whole core accident) 全堆芯事故
WDCS (wet-well dry-well cooling system) 湿井干井冷却系
WDR (water displacer rod) 挤水棒
weber (Wb) 韦伯(磁通量单位)
wet bulb thermometer 湿球温度计
wet-well dry-well cooling system (WDCS) 湿井干井冷却系
Wheatstone bridge 惠斯通电桥
wheel efficiency 轮周效率
whole-body counter (WBC) 全身计数器
whole-body X-irradiation X 射线全身辐照
whole core accident (WCA) 全堆芯事故
wide-band data (WBD) 宽(频)带数据
wide-band data link (WBDL) 宽(频)带数据链路
windage loss 风耗,空气摩擦损耗
winding 绕组
winding bar 线棒
winding coil 绕组线圈
winding factor 绕组因数
winding pitch 绕组节距
winding with graded insulation 分级绝缘绕组
wind load 风荷载
wind turbine generator (WTG) 风电机,风力发电机组
wing wall 翼墙
wire enamel 漆包线漆
wire flattening and profiling machine 型线轧拉机
withdrawable part 可抽件
work breakdown structure (WBS) 作业分解结构
working earthing 工作接地
working head 工作头
working medium 工质
working voltage (WV) 工作电压
working space 工作空间
work storage 工作存储器
wound core 卷绕铁芯
wrapping separation 包护隔离
write buffer (WB) 写缓冲器
wrought aluminium alloy 变形铝合金
WTG (wind turbine generator) 风电机,风力发电机组
W-type fibre W-型光纤
Wulf electrometer 乌尔夫静电计
wustite 方铁矿
WV (working voltage) 工作电压
wye Y 形支架
wye-delta 星形-三角形连接(法)
wye-delta power transformer 星形-三角形电力变压器
wye voltage 星形电压,Y 接法电压

x-axis x轴
x-axis amplifier x轴信号放大器
X band X频带
X-brace 叉梁,交叉支撑
X-bridge X型电桥,电抗电桥
x-coordinate 横坐标,x坐标
x-direction x轴方向
Xe and Sm poisoning 氙和钐中毒
Xe-bank position control loop 氙控制组位置控制回路
xenon accumulation 氙积累
xenon build-up 氙积累
xenon oscillation 氙振荡
xenon poisoning 氙中毒
Xe-rod bank 氙棒组
xerogel 干凝胶
xerographic printer 静电印刷机
xerography 静电印刷术

xeroradiography 干板X射线照相术
X-network X形网络
X-ray diffraction (XRD) X射线衍射
X-ray fluoroscope (XRF) X射线荧光检查仪
X-ray laser (XRL) X射线激光器
X-ray machine X射线机
X-ray tube current X射线管电流
XRD (X-ray diffraction) X射线衍射
XRF (X-ray fluoroscope) X射线荧光检查仪
XRL (X-ray laser) X射线激光器
X-wax X蜡
X-Y recorder X-Y记录器

y-axis y 轴,纵轴
y-axis amplifier y 轴线信号放大器
Y-bend 裤衩管
Y box 星形连接电阻箱,Y 形连接电阻箱
Y circulator Y 环行器
Y-class insulation Y 级绝缘材料
Y-connection 星形联结,Y 接法
Y-connection phase winding 星形接法相绕组,Y 接法相绕组
y-coordinate 纵坐标
Y-cut Y 切割
Y-delta starter 星形-三角形起动器
y-direction y 轴方向
year-end adjustment 年终调整
yearly efficiency 年效率
yearly load curve 年负荷曲线,年负载曲线
yearly load factor 年负荷率,年负载率
yearly maximum load 年最高负荷,年最高负载
yes-no decision 是非决策
yield criterion 屈服准则
yield cross-section (裂变产物)产额截面
yield curve (裂变)产额曲线
yield distribution 产额分布
yield factor 产额因子
yield fraction 产额份额
yield point 屈服点
yield strength 屈服强度
yield stress 屈服应力
Y-jet atomizer Y 型雾化器
Y-jet type oil burner Y 型油喷燃器
Y joint Y 型接头
Y-junction 星形联结;Y 接头
Y-matching Y 形匹配
YME (Young's modulus of elasticity) 杨氏弹性模量
yoke 铁轭,磁轭
yoke air gap 轭空气隙
yoke ampere-turns 轭安匝
yoke flux 轭磁通
yolk coal 无黏性煤
Young's modulus 杨氏模量,杨氏模数
Young's modulus of elasticity (YME) 杨氏弹性模量
Y-stay Y 形拉线
Y-type core Y 型铁芯
Y-voltage (星形联结中的)相电压
Y-Y connection 双星形接法

Z-angle Z型角铁
z-domain z域
Zener breakdown 齐纳击穿
zeolite 沸石
zero 零点
zero adjustment 调零,零位调整
zero-based conformity 零基一致性
zero-based linearity 零基线性度
zero calibration gas 零点校准气
zero capacitance 零电容
zero discharge 零排放
zero displacement value 零位位移值
zero drift 零(点)漂移
zero elevation 零点提升
zero error 零点误差
zero gas 零气
zero inductance 零电感
zero-input response 零输入响应
zero-length spring gravimeter 零长弹簧重力仪
zero offset 零点偏移
zero output 零输出
zero output baseline 零输出基线
zero-phase-sequence relay 零序继电器
zero point 零点
zero-point output 零点输出
zero-sequence component 零序分量
zero-sequence current protection 零序电流保护
zero-state response 零状态响应
zigzag connection Z形联结,曲折形联结
zinc 锌
zinc-carbon battery 碳锌电池
zincification 镀锌
zircaloy 锆锡合金
zirconate 锆酸盐
zonation 分带,分区
zoned air supply 分段送风
zoogloea 菌胶团
zoom 缩放
Z-section steel Z型钢
Z-transform Z变换

汉英部分

阿尔法射线 alpha radiation
爱迪生效应 Edison effect
爱普斯坦方圈 Epstein frame
安培表 ampere-meter
安培小时 Ah (ampere-hour)
安全棒 safety rod
安全冲洗距离 safety washing distance
安全电路 safety circuit
安全阀 SV (safety valve)
安全壳 containment
安全联锁系统 safety interlock system
安全逻辑装置 safe logic assembly
安全停堆 safe shutdown
安全裕度 margin of safety
安全制动器 fail-safe brake
安全注射箱 accumulator
安时 Ah (ampere-hour)
安装 fitting; installation
安装成本 installation cost
安装互换性 intermountability
安装间距 installation clearance
氨 ammonia
氨化作用 ammonification
氨量计 ammoniometer
氨水 aqua ammonia
铵 ammonium
按键开关 key switch
暗装布线 installation under plaster; installation under the surface
盎司 ounce
凹槽 notch
凹槽衬垫 pit liner
凹角 reentrant
凹口 notch
凹模 matrix
奥罗管 orotron
奥斯特 Oe (oersted)

八极管 octode
八进制计数法 octal notation
八进制数字 octal digit
巴 bar
靶 target
靶核 target nucleus
靶截止电压 target cut-off voltage
靶区 target area
靶式流量计 target flowmeter
白炽灯 incandescent light
柏油环氧漆 tar epoxy paint
摆动 oscillation
摆动喷嘴 tilting nozzle
摆动(式)燃烧器 tilting burner
摆度 throw
摆环式转速计 tilting-ring tachometer
摆线 trochoid
扳手 spanner
板材 sheet
板极 plate
板极检波器 transrectifier
板式换热器 plate heat exchanger
板式基础 slab foundation
板条 batten; lath
板岩 slate
板状燃料组件 plate fuel assembly
半波 loop
半波整流器 half-wave rectifier
半导体开关装置 semiconductor switch device
半导体样品台 semiconducting specimen stage
半导体应变计 semiconductor strain gauge
半导体振荡器 oscillistor
半电波暗室 semi-anechoic chamber
半电桥 half-bridge
半加(法)器 two-input adder
半减(法)器 two-input subtracter
半径 radius
半控制电桥 half-controlled bridge
半模拟盘 semi-graphic panel
半模拟屏 semi-graphic panel
半圈式绕组 half-coiled winding
半实物仿真 semi-physical simulation
半数字读出 semidigital readout
半衰期 half-period
半双工接口 half-duplex interface
半透明体 translucent body
半线圈 bar
半周期 half-period
半自动测井仪 semi-automatic logging instrument
半自动化 semi-automation
半自动试验机 semi-automatic testing machine
伴随故障 incidental defect
棒式绝缘子 strut insulator
包层铝 Alclad
包带机 taping machine
包方式 packet mode
包护隔离 wrapping separation
包交换 packet switching
包壳温度 clad temperature
包排序 packet sequencing

胞状组织　cellular structure
薄板　lamella; lamina; leaf; sheet
薄层　lamina
薄片　lamella; leaf
饱和　saturation
饱和差　saturation deficit
饱和电抗器　transductor
饱和甘汞电极　saturated calomel electrode
饱和式标准电池　saturated standard cell
饱和水汽压　saturation vapour pressure
饱和特性　saturation characteristics
饱和蒸汽极限回热循环　ultimate regenerative cycle of saturated steam
保安性　fail-safe
保持操作　lockout operation
保持继电器　latching relay; lockout relay
保护电流互感器　protective current transformer
保护环　guard ring
保护爬电距离　protected creepage distance
保护输入　guarded input
保护因数　protection factor
保护裕度　protective margin
保护罩　guard
保护中性导体　PEN conductor
保护装置　guard
保活　keep-alive
保温损耗　standing loss
保险丝　fuse
保证电能　firm energy
报警灯　warning light
报警电路　warning circuit
报警指示信号　AIS (alarm indication signal)
报警装置　warning device
抱杆　derrick
爆击　knocking
爆燃　deflagration; detonation; knocking combustion
爆震性　knock property
爆震性试验　knock test
贝塞尔函数表值　tabular values of Bessel function
备用容量　margin capacity
背板　shuttering
钡　barium
倍频　multiple frequency
倍增管　multiplier tube
倍增器　multiplier
倍增器增益　multiplier gain
锛子　adze
本安型　intrinsic safety
本地电池　LB (local battery)
本地环路　local loop
本机反馈　LFB (local feedback)
本机振荡器　local oscillator
本身消耗　intrinsic consumption
本振频率　local frequency
本征半导体　intrinsic semiconductor
本征变位比　intrinsic stand-off ratio
本征导电　intrinsic conduction
本征导电率　intrinsic conductivity
本征电导率　intrinsic conductivity
本征感应　intrinsic induction
本征结型晶体管　intrinsic junction transistor
本征抗电强度　intrinsic electric strength
本征频率　eigenfrequency
本征值　eigenvalue
本质安全电路　intrinsically safe circuit
本质安全位垒　intrinsic safety barrier
本质安全型　intrinsic safety
本质安全型电气设备　intrinsically safe electrical apparatus
本质失效　inherent weakness failure

苯 benzene
泵 pump
比例控制 proportional control
比例控制器 proportional controller
比例作用 P-action
比容 specific volume
比色法 colorimetry
比色计 colorimeter
比特 bit
比体积 specific volume
比重计 densimeter
比转速 specific speed
闭环控制 feedback control
闭环控制系统 CLCS (closed loop control system)
闭开触点 make-and-break contact
闭式循环冷却水 CCCW (closed cycle cooling water)
闭锁电磁铁 latching electromagnet
闭锁电流 latching current
闭锁继电器 latched relay
避雷器 lightning arrester
避雷线 ground wire; lightning conductor
避雷针 lightning conductor; lightning rod
臂 arm
边发射 edge emission
边际频数 marginal frequency
边界频率 edge frequency
边界线 BL (boundary line)
边界效应 interface effect
边缘不规则性 edge irregularity
边缘发射 edge emission
边缘效应 edge effect
扁斧 adze
扁率 oblateness
扁绕机 strip-on-edge winding machine
扁圆形 oblateness
变比 transformation ratio
变比计 transformation-ratio meter
变磁性 metamagnetism
变磁阻式传感器 variable reluctance transducer
变电所 transformer station; (transformer) substation; transforming station
变电站 transformer station; (transformer) substation; transforming station
变感器 variometer
变工况运行 variable parameter operation
变化器 variator
变化压力 variable pressure
变换方程 transformation equation
变换器 transformer
变换群 transformation group
变结构控制系统 variable structure control system
变截面流量计 variable-area flowmeter
变量 variable
变流比 current transformation ratio; transformer ratio
变流变压器 converter transformer
变流器 converter
变面积法 variable-area method
变面积流量计 variable-area flowmeter
变扭器传动 torque converter transmission
变容二极管 varactor; variable capacitance diode; varicap
变像管 image converter tube
变形铝合金 wrought aluminium alloy
变压 variable pressure
变压比 transformation ratio; transformer ratio
变压比电桥 transformer-ratio bridge
变压器 transformer
变压器差动保护 transformer differential protection
变压器冲片 transformer stamping

变压器磁轭 transformer yoke
变压器磁分路 transformer magnetic shunt
变压器等效电路 transformer equivalent circuit
变压器电动势 transformer electromotive force
变压器电抗 transformer reactance
变压器电压 transformer voltage
变压器断路器 transformer breaker
变压器分接头 transformer tap
变压器硅钢 transformer steel
变压器硅钢片 transformer sheet
变压器继电器 transformer relay
变压器馈线保护 transformer feeder protection
变压器冷却器 transformer cooler
变压器利用率 transformer utilization factor
变压器利用系数 transformer utilization factor
变压器耦合 transformer coupling
变压器耦合放大器 transformer-coupled amplifier
变压器耦合负荷 transformer-coupled load
变压器耦合脉冲放大器 transformer-coupled pulse amplifier
变压器耦合振荡器 transformer-coupled oscillator
变压器匹配 transformer matching
变压器绕组 transformer winding
变压器室 transformer cabin; transformer vault
变压器输出 transformer output
变压器塔柱 transformer pillar
变压器套管 transformer bushing
变压器铁芯 transformer core
变压器铁芯柱 leg
变压器外壳 transformer casing
变压器网络 transformer network
变压器线圈 transformer coil
变压器箱 transformer tank
变压器效应 transformer effect
变压器压降补偿器 transformer drop compensator
变压器引线 transformer lead
变压器油 transformer oil
变压器油冷却器 transformer oil cooler
变压器匝(数)比 transformer turn ratio
变压器转子 transformer rotor
变压器组 transformer bank
变压器作用 transformer action
变压头流量计 variable-head flowmeter
变异 mutation
变址寄存器 index register
变转速风轮 variable speed rotor
辨识 identification
标称电流 nominal current
标称电压 nominal voltage
标称电阻 nominal resistance
标称频带 nominal band
标称容量 nominal capacity
标称值 nominal value
标尺分格值 value of scale division
标定 calibration
标定比 marked ratio; nominal ratio
标度值 value of scale mark
标高 elevation
标号编码 label coding
标号常数 label constant
标号数据 label data
标记电路 marking circuit
标记频率 mark frequency
标识符 identifier
标识器 marker
标准电池允许放电量 permissible discharge of standard cell
标准电池允许累计放电量 permissible cumulative discharge of standard cell

标准贯入试验 SPT (standard penetration test)
标准化 normalization; standardization
标准化电阻 normalized resistance
标准化频率 normalized frequency
标准化曲线 normalized curve
标准煤 standard coal
标准软铜 standard annealed copper
标准正交性 orthonormality
标准阻抗 normal impedance
表 meter
表层物 skim
表格显示(器) tabular display
表解 tabulated solution
表面式加热器 surface-type heater
表驱动编译程序 table-driven compiler
表土 topsoil
表值 tabular value
丙烯腈-丁二烯-苯乙烯 ABS (acrylonitrile-butadiene-styrene)
并联 parallel connection; paralleling
并联电路 parallel circuit
并联电容补偿 parallel capacitive compensation
并联运行 multiple operation; paralleling; parallel operation
并行操作 parallel operation
并行计算机 parallel computer
并行输出制 parallel output system
并行性 parallelism
拨动开关 toggle switch
拨动式开关电容 toggle switched capacitor
波瓣 lobe
波瓣频率 lobe frequency
波包 wave packet
波长 wavelength

波长常数 wave constant
波长计 cymometer; wavemeter
波长转换开关 wave changing switch
波程角 wave angle
波传播 wave propagation
波导 wave director; waveguide
波导磁控管 waveguide magnetron
波导存储器 waveguide storage
波导反射器 waveguide reflector
波导分支滤波器 waveguide branching filter
波导管 waveguide
波导结 waveguide junction
波导开关 waveguide switch
波导滤波器 waveguide filter
波导调制器 waveguide modulator
波导通信 waveguide communication
波导型调速管 waveguide klystron
波导阻抗 waveguide impedance
波道 wave channel
波电压 wave voltage
波顶 wave crest
波动 wave motion
波动过程 wave process
波动频率 wave frequency
波动声学 wave acoustics
波动说 wave theory
波动微分方程 wave differential equation
波段 band; wave range
波段开关 waver; wave-range switch; wave selector
波段转换开关 wave changing switch
波发生器 wave generator
波法线 wave normal
波峰 wave crest; wave ridge
波辐射 wave radiation
波腹 wave loop
波干扰 wave interference
波干涉 wave interference

波谷　wave trough
波函数　wave function
波加热器　wave heater
波节　knot; wave node
波浪电站　wave-power station
波浪发电　wave-power generation
波浪发电装置　wave-power device
波浪反射　wave reflection
波力电机　wave motor
波力电站　wave-power station
波力发电　wave-power generation
波列　wave train
波列频率　wave-train frequency
波轮　impeller
波模　wave mode
波能　wave energy
波能发电机　wave energy generator
波频　wave frequency
波频计　wavemeter
波谱　wave spectrum
波谱分辨率　spectral resolution
波前　wavefront
波前匹配　wavefront matching
波前倾角　wavefront angle
波前倾斜　wave tilt
波群　wave group
波绕线圈　wave coil
波绕组　wave winding
波矢量　wave vector
波数　wave number
波衰减　wave attenuation
波速　wave velocity
波特　baud
波头截断冲击波　impulse chopped on the front
波尾　wave tail
波尾截断冲击波　impulse chopped on the tail
波纹(钢)板　corrugated steel sheet
波线　wave line
波形　waveform; waveshape

波形变换器　wave converter; waveform converter
波形分离　waveform separation
波形分析器　wave analyzer
波形合成器　waveform synthesizer
波形畸变　waveform distortion
波形记录器　waveform recorder
波形脉冲　waveform pulse
波形区分　waveform separation
波形失真　klirr; waveform distortion
波形试验　waveform test
波形图　oscillogram
波形形成　waveshaping
波形形成器　waveshaper
波形因数　form factor
波形转换器　waver
波形自记器　kymograph
波型　wave pattern
波型选择器　wave selector
波型振动　wave-type vibration
波阵面　wavefront
波阵面匹配　wavefront matching
波阵面倾角　wavefront angle
波状流　wavy flow
波状运动　wave motion
波阻　wave drag
波阻抗　wave impedance
波阻力　wave drag
玻璃纤维　spun glass
玻璃纤维增强塑料　FRP (fibreglass-reinforced plastic)
泊　P (poise)
博弈论　game theory
箔　leaf
箔式线圈　foil coil
箔验电计　leaf electrometer
箔验电器　leaf electroscope
薄膜电容器　film capacitor
薄膜型掩模　membrane mask
补偿　offset
补偿棒　shim rod
补偿电流　offset current
补偿电压　offset voltage
补机推送　pusher operation

补燃 afterburning
补燃室 afterburner
捕收剂 collector
捕捉器 arrester; catcher
不导通间隔 idle interval
不动作值 non-operating value
不断路星形-三角形起动器 Wauchope type starter
不对称度 asymmetry
不腐蚀性 non-corrodibility; non-corrosibility
不复位开关 non-homing switch
不间断电源 UPS (uninterrupted power supply)
不接地 isolation from earth
不接地电压互感器 unearthed voltage transformer
不接地电源 isolated supply
不接地中线 isolated neural
不均匀场 non-uniform field
不可拆件 non-detachable part
不可重接插头 non-rewirable plug
不可用系数 UF (unavailable factor)
不可用小时 UH (unavailable hours)
不灵敏性 insensitivity
不漏电的 leakproof
不敏感性 insensitivity
不平衡电流 out-of-balance current; unsymmetrical current
不平衡度 asymmetry
不平行性 non-parallelism
不确定故障 indeterminate fault
不燃结构 non-combustible construction
不守恒 nonconservation
不同轴性 non-axiality
不透明等离子体 opaque plasma
不透明度 opacity
不透明区 opaque region
不透水密封 watertight seal
不透水外壳 watertight closure
不透水性 watertightness
不完全桥式连接 incomplete bridge connection
不吸动值 non-pickup
不锈钢 corrosion-resistant steel; SS (stainless steel)
不一致 non-coincidence
布局 layout; placement
布局图 layout chart
布氏硬度数 BHN (Brinell hardness number)
布置 layout
布置数据 layout data
布置图 layout plan
步进电动机 stepping motor
步进式开关 stepping switch
步距 step pitch
部分导体短路 inter-stand short circuit
部件 sub-assembly

采石厂 quarry
采样保持器 sampling holder
采样横截面 sampling cross-section
采样间隔 sampling interval
采样控制 sampling control
采样控制器 sampling controller
采样控制系统 sampling control system
采样脉冲 sampling pulse
采样频率 sampling frequency
采样时间 sampling time
采样速率 sampling rate
采样误差 sampling error
采样系统 sampling system
采样元件 sampling element
采样值 sampling value
采样周期 sampling period
采样作用 sampling action
参比电极 reference electrode
参比记录纸 reference chart
参比接点 reference junction
参比条件 reference condition
参比柱 reference column
参考试块 reference block
参考线圈 reference coil
参考样本 reference specimen
参量 parameter
参数 parameter
残差 residual error
残错率 residual error rate
残余标准偏差 residual standard deviation
残余电流 residual current
残余方差 residual variance
残余偏转 residual deflection
残余载波 residual carrier

残渣 tailing(s)
残渣油 residual fuel oil
操动件 actuator
操纵杆 joystick
操纵台 console
操作 operation
操作冲击试验 switching impulse test
操作冲击水平 SIL (switching impulse level)
操作过电压 switching overvoltage
操作码 operation code
操作说明 instruction
槽 gutter; slot
槽电压 tank voltage
槽路 tank circuit
槽路电容器 tank capacitor
槽轮 sheave
槽楔 slot wedge
槽型反应器 tank reactor
侧向弯度 camber
测波杆 wave staff
测尘器 konimeter
测电法 electrometry
测电术 electrometry
测电学 electrometrics
测光仪 photometer
测角器 clinometer
测控 instrumentation and control
测量电桥 measuring bridge
测量放大器 instrumentation amplifier
测偏仪 derivometer
测试设备 TE (test equipment)
测试与置位 TAS (test and set)

测试运行　TR（test run）
测速发电机　tachogenerator; tachometer generator
测速发送机　tachometer sender
测速热敏电阻器　velocity measuring thermistor
测压管　piezometer
测压计　piezometer
层　bed
层错　fault
层间电压　interlayer voltage
层间短路　inter-lamination fault
层间绝缘　interlayer insulation; layer insulation
层间连接　interlayer connection
层间信号串扰　layer-to-layer signal transfer
层流　laminar flow
层流等离子枪　laminar plasma torch
层式电缆　layered cable
层式线圈　layer coil
层析X射线照相机　tomograph
层析X射线照相术　tomography
层压板　laminate
层压管　laminated tube
层状淀积　laminar deposition
叉梁　X-brace
差　differential
差压　differential pressure
插槽　slot
插点　intermediate point
插孔　jack
插孔板　jack panel
插口　jack
插入拔出力　insertion and withdrawal force
插入半径　inserted radius
插入电流　insertion current
插入绕组　push-through winding
插入式断路器　plug-in circuit-breaker
插入物　inlet
插套　sleeve
插头〖带保护接地触点的〗　grounding-type plug
插头式电阻箱　resistance box with plug
插销　bolt; pin
插针　pin
插座　socket（outlet）
插座〖带保护接地触点的〗　grounding-type receptacle
插座配电盘　jack distributor
掺和料　admixture
掺和物　admixture
掺气　AE（air entraining）
产额分布　yield distribution
产额份额　yield fraction
产额截面　yield cross-section
产额曲线　yield curve
产额因子　yield factor
产形齿轮　generating gear
铲　spade
铲运机　scraper
长臂圆规　beam-compasses
长度　l（length）
长度系数　length coefficient
长距绕组　long-pitch winding
常量　constant
常数　constant
常态　normality
常项　constant
厂内系统　intra-plant system
厂用变压器　house transformer
厂用发电机　house generator
场　field
场结构　field structure
场调管　technetron
场效应高能晶体管　technetron
超钚　transplutonium
超测微计　ultramicrometer
超纯度　ultra-purity
超纯水　ultrapure water
超导性　superconductivity
超低频　ULF（ultra-low frequency）
超低温热敏电阻（器）　ultra-low-temperature thermistor
超低污染排放　ultra-low emission
超电势　overpotential
超短波　ultrashort wave

超镄元素 transfermium element
超复励 overcompounding
超高纯度 ultra-high purity
超高纯度氦气氛(层) ultra-high-purity helium atmosphere
超高分辨力 ultra-high resolution
超高频 UHF (ultra-high frequency)
超高频陶瓷电容器 UHF ceramic capacitor
超高频调谐器 ultra-high-frequency tuner
超高速瞬时波形分析器 ultra-high-speed transient waveform analyzer
超高温 UHT (ultra-high temperature)
超高温反应堆 ultra-high-temperature reactor
超高压 EHT (extra-high tension); EHV (extra-high voltage)
超高压放电 extra-high pressure discharge
超高真空 extra-high vacuum; UHV (ultra-high vacuum)
超过滤 ultrafiltration
超过限度 overrun
超净空气系统 ultra-clean air system
超快中子 ultrafast neutron
超铹元素 translawrencium element
超冷中子 ultracold neutron
超临界机组 supercritical plant
超灵敏检漏法 ultra-sensitive leak testing method
超流体 superfluid
超滤 ultrafiltration
超滤法 ultrafiltration process
超滤膜 ultrafiltration membrane
超滤膜法 ultrafiltration membrane process
超滤器 ultrafilter
超滤系统 ultrafiltration system
超滤液 ultrafiltrate
超铌元素 transnobelium element
超前 lead
超前补偿 lead compensation
超前电流 leading current
超前相 leading phase
超前滞后 lead-lag
超前组件 lead module
超声波报警系统 ultrasonic alarm system
超声波雾化器 ultrasonic atomizer
超声频率 ultra-audible frequency
超声束 ultrasonic beam
超声学 ultrasonics
超声振荡器 ultrasonator
超速离心机 ultracentrifuge
超钍元素 transthorium element
超微分析 ultra-microanalysis
超微颗粒 ultrafine particle
超微离子交换树脂 ultrafine ion-exchange resin
超微量测定 ultramicro-determination
超微量分析 ultra-microanalysis
超微量化学 ultramicrochemistry
超微量天平 ultramicrobalance
超显微镜 ultramicroscope
超谐波 ultraharmonic
超音频 ultra-audible frequency
超铀元素 transuranium element
超载 overload
超真空阀 ultra-high vacuum valve
潮波 tidal wave
潮高基准面 tidal datum
潮流 tidal current
潮汐波 tidal wave
潮汐电站 tidal power station
潮汐动力 tidal power
潮汐发电 tidal power generation
潮汐发电机 tidal power generator
潮汐发动机 tide-motor
潮汐能 tidal energy; tidal power
车间组装制作件 SAFP (shop-assembled fabricated piece)

沉积(物) sediment
沉降 subsidence
沉陷 subsidence
沉渣 sludge
衬垫 cushion; pad
撑脚 arm-brace
成层 lamination
成核剂 nucleater
成列直插封装开关 in-line package switch
成套装置隔离 isolation of a unit
成像系统 imaging system
成形绕组 transfer winding
承拉螺栓 tie bolt
承载能力极限状态 ultimate limit state
承载时间 in-service period
城市电网规划 urban power network planning
乘法电路 multiplication circuit
乘法器 multiplier
乘数 multiplier
澄清器 settler
弛度 sag
驰振 gallop
持久屈服极限 time yield limit
持平值 holding value
持续短路试验 heat run
持续时间 duration
持续性 continuity
齿 tooth
齿波纹 tooth ripple
齿部饱和 tooth saturation
齿顶高 addendum
齿尖漏抗 tooth-tip reactance
齿距 tooth pitch
齿式联轴器 tooth coupling
齿条 toothed bar
齿谐波 tooth harmonic
齿形波纹 tooth ripple
冲程 stroke
冲管 pipeline flushing
冲击成型 impact moulding
冲击传感器 shock sensor; shock transducer
冲击电流 impulse current
冲击电流限制器 impulse-current limiter
冲击电流最大值 inrush peak
冲击电压 impulse voltage
冲击放电试验 impulse sparkover test
冲击峰值电压 impulse crest voltage
冲击负荷 impulse load
冲击负载 impulse load
冲击击穿 impulse breakdown
冲击击穿强度 impulse breakdown strength
冲击激励 shock excitation
冲击力矩 impulse torque
冲击率 jerk rate
冲击耐受电压 impulse withstand voltage
冲击强度 impact strength
冲击闪络 impulse flashover
冲击闪络试验 impulse flashover test
冲击试验 impact test; impulse test; shock test
冲击试验波尾电压 tail-of-wave impulse test voltage
冲击试验机 impact testing machine
冲击试验设备 impact testing machine
冲击试验台 shock testing machine
冲击试验站 impulse testing station
冲击台 shock testing machine
冲击限制 jerk limitation
冲击响应谱 shock response spectrum
冲击抑制电路断路器 inrush suppressor circuit-breaker
冲击应力 impact stress
冲击振动 shock vibration
冲积层 alluvium
冲角损失 incidence loss
冲蚀度 abrasivity
冲压 stamping

冲压件 stamping
充电 charge
充气 aeration
充气器 aerator
充水裂缝 water-filled fracture
重沸器 reboiler
重复接地 iterative earthing
重复阻抗 iterative impedance
重接通时间 re-make time
重碳酸盐 bicarbonate
重言式 tautology
重影 echo-image
重组 realignment
抽汽口 tapping point
抽取 abstraction
抽水蓄能电站 pumped storage station
抽头 tap
抽头电压 tap voltage
抽头定子绕组 tapped stator winding
抽头感应线圈 tapped induction coil
抽头换接开关 tap changer
抽头切换 tap changing
抽头切换变压器 tap-change transformer; tap-changing transformer
抽头式变压器 tapped transformer
抽头调谐电路 tapped tuned circuit
抽头线 tapped line
抽头线圈 tapped coil
抽头延迟线 tapped delay line
抽头匝数比 tapping ratio
抽象(化) abstraction
抽样试验 sampling test
稠密栅格 tight lattice; tight-packed lattice
臭氧 ozone
臭氧发生器 ozone generator; ozonizer
出口边损失 trailing section loss
出料槽 spout
出水口 water outlet

出线走廊 outlet line corridor
初电容 initial capacitance
初级绕组 primary winding
初始电压 initial voltage
初始断口 incipient break
初始裂缝 incipient crack
初始裂纹 incipient crack
初始破裂 incipient break
初始位置 home position
初始应力 initial stress
初始转矩 initial torque
初应变 initial strain
初应力 initial stress
除气器 degasifier
除湿器 dehumidifier
除盐 demineralization
除氧器 deaerator
处理单元 processor
处理机 processor
处理器 processor
储备电池 reserve cell
储热层 reservoir
储水系数 storativity
触点断开 off contact
触电 electric shock
触发电路 toggle circuit
穿晶断裂 transgranular fracture
穿晶腐蚀 transgranular corrosion
穿晶破裂 transgranular cracking
穿通式套管 throughway bushing
穿透 penetration
穿透式半导体探测器 transmission semiconductor detector
穿心电容器 feed-through capacitor
传递比 transfer ratio
传递导抗 transfer immittance
传递函数放大器 transfer function amplifier
传递函数分析器 transfer function analyzer
传递机构 transfer mechanism
传递面 transfer surface
传递式换热器 transfer-type heat exchanger

传递通路 transmission path
传递因数 transfer factor
传递阻抗 transfer impedance
传动 drive
传动比 transmission ratio
传动机构 transfer mechanism
传动式测力计 transmission dynamometer
传动轴 jack shaft
传感器 pickup; sensor; transducer
传热系数 heat transfer coefficient; thermal transmission coefficient
传输比 transmission ratio
传输常数 transfer constant
传输当量 transmission equivalent
传输电平 transmission level
传输范围 transmission range
传输接口转换器 transmission interface converter
传输控制 transmission control
传输控制器 TC (transmission controller)
传输控制协议 TCP (transmission control protocol)
传输控制元件 TCE (transmission control element)
传输路径 transmission path
传输路线 transmission route
传输路由 transmission route
传输脉冲 transmission pulse
传输媒体 transmission medium
传输媒质 transmission medium
传输频率 transmission frequency
传输衰减 transmission attenuation
传输速率 transmission rate
传输调节器 transmission regulator
传输通道 transmission path
传输通路 transmission channel
传输系数 transmission coefficient; transmission factor
传输线 transmission line
传输协议 transmission protocol
传输性能 transmission performance
传输延迟 transmission delay
传输滞后 transmission lag
传输转接器 transmission adaptor
传输子系统接口 transmission subsystem interface
传送变阻器 transmitting rheostat
传送控制寄存器 transfer control register
传送容器 transfer flask
传质面 transfer surface
船闸 ship lock
串并联电路 series-parallel circuit
串并行输出制 series-parallel output system
串级磁镜反应堆 tandem mirror reactor
串级磁镜聚变裂变混合堆 tandem mirror fusion-fission hybrid
串级磁镜试验 tandem mirror experiment
串联 series connection; tandem connection
串联刀开关 tandem knife switch
串联电感器 series inductor
串联电路 series circuit
串联电容器 series capacitor
串联电阻器 series resistor
串联发电机 tandem generator
串联控制 tandem control
串联配置 tandem arrangement
串联起电器 tandem generator
串联式水泵 tandem type pump
串联运行 series operation
串列式发动机 tandem engine
串列式叶片 tandem blade
串模电压 series-mode voltage
串模信号 series-mode signal
串模抑制 series-mode rejection
串模抑制比 SMRR (series-mode rejection ratio)

串扰 crosstalk
串行操作 serial operation
串行处理 serial processing
串行传输 serial transmission
串行存取 serial access
串行计算机 serial computer
串行输出制 series output system
串音 crosstalk
吹灰器 soot blower
吹氧 oxygen blow
垂度 sag
垂极 orthopole
垂向扫描器 orthoscanner
垂心 orthocentre
垂直极限 vertical limit
垂直角 vertical angle
垂直能见度 vertical visibility
垂直扫描 vertical scanning
垂直尾翼 vertical tail
垂直轴 vertical axis
锤式打桩机 monkey (engine)
瓷夹板 porcelain cleat
瓷绝缘子 porcelain insulator
瓷套管 porcelain bushing
磁饱和 magnetic saturation
磁变管 magnistor
磁变仪 variometer
磁波子 magnon
磁场放大机 metadyne
磁场线圈 field coil
磁带操作系统 TOS (tape operating system)
磁带绘图系统 tape plotting system
磁导计 permeameter
磁导率 permeability
磁电系检流计 permanent-magnet moving coil galvanometer
磁电系仪表 permanent-magnet moving coil instrument
磁动势 MMF (magnetomotive force)
磁轭 yoke
磁放大器 magnetrol; transductor (amplifier)

磁放大器控制发生器 transductor-controlled generator
磁分路补偿 magnetic shunt compensation
磁粉 magnetic particle
磁刚度 magnetic rigidity
磁光学 magneto-optics
磁化率 susceptibility
磁化器 magnetizer
磁极 magnetic pole
磁开关 magnistor
磁控等离子体开关 madistor
磁控管 magnetron
磁力 magnetism
磁敏电阻(器) magnetoresistor
磁敏二极管 magneto-diode
磁摩擦离合器 magnetic friction clutch
磁扭线 twistor
磁屏蔽仪表 instrument with magnetic screen
磁强计 magnetometer
磁强记录仪 magnetograph
磁石发电机 magneto
磁体 magnet
磁通(量) induction flux; magnetic flux
磁通势 MMF (magnetomotive force)
磁图 magnetogram
磁性 magnetism
磁学 magnetics
磁致电阻 magnetoresistance
磁致电阻器 magnetoresistor
磁致伸缩 magnetostriction
磁滞损耗 hysteresis loss
磁子 magneton
磁阻 magnetoresistance; reluctance
磁阻器 magnetoresistor
次级电子倍增 multipacting
次级绕组 secondary winding
次末级 penultimate stage
次烟煤 sub-bituminous coal
刺墙 key wall

从动带轮　driven pulley
粗粉分离器　classifier
粗滤器　strainer
猝灭剂　quencher
猝灭效应　quenching effect
猝熄　quenching
猝熄变压器　quenching transformer
猝熄电路　quenching circuit
猝熄火花隙　quenched spark gap
猝熄频率　quench frequency
醋酸丁酸纤维素　cab (cellulose acetate butyrate)
催化　catalysis
催化剂　catalyst
脆性　brittleness
淬火　quenching
淬火冷却槽　quenching tank
淬火裂纹　quench crack
存储程序控制　SPC (stored-program control)
存储管　storage tube
存储寄存器　memory register
存储介质　storage medium
存储器　accumulator
存储容量　memory capacity
存储示波器　storage oscilloscope
存储芯片　memory chip
错峰　staggering peak load

搭焊 joint welding
搭接 lap-over
搭接焊 lap welding
搭接接头 lap joint
达因 dyne
打夯机 tamper
大半波 major loop
大地电导率 earth conductivity
大地电流 EC (earth current); ground current
大地电位 ground potential
大地电阻 earth resistance; ground resistance
大地电阻率 earth resistivity
大地回路 ER (earth return)
大电流 heavy current
大电流汇流排 heavy-current bus
大电流连接器 heavy-current connector
大电流脉冲 high-current impulse
大电流母线 heavy-current bus
大电子显示器 LED (large electronic display)
大分子 macromolecule
大功率变压器 high-rating transformer
大功率电动机 high-power motor
大功率断路器 heavy-duty circuit-breaker
大功率放大器 high-power amplifier
大功率综合电路 high-power synthetic circuit
大规模集成电路 LSI (large-scale integration)
大卡 kcal (kilocalorie)
大气压力 barometric pressure
大切断功率熔断器 high-breaking-capacity fuse; high-interrupting-capacity fuse
大头钉 stud
大修间隔 TBO (time between overhauls)
大中型直流电动机 large and medium DC motor
代表性过电压 representative overvoltage
代码 code
代码转换器 code converter
带泵加热器 pumped heater
带比较器 tape comparator
带标识器 tape identification unit
带操作系统 TOS (tape operating system)
带电拔除插头 hot unplugging
带电部分 live part
带电插入 hot insertion
带电电路 live circuit
带电电线 hot wire
带电线路 live line
带发送分配器 tape transmitter-distributer
带负荷试验 on-load test
带鼓 tape drum
带环存储器 tape-loop storage
带奇偶 tape parity
带接头电刷 headed brush
带绝缘 tape insulation
带绝缘层 tape covering
带绝缘线圈 tape insulated coil
带控装置 tape control unit
带绕磁芯 tape-wound core

带绕绝缘 taped insulation
带熔断器开关 switch-fuse
带限(制) tape limit
带引导例程 tape bootstrap routine
带准备装置 tape preparation unit
戴维南定理 Thevenin theorem
单波绕组 simplex wave winding
单刀单掷 SPST (single-pole, single-throw)
单刀双掷 SPDT (single pole, double throw)
单地址码 one-address code
单电动机 monomotor
单电子 lone electron
单叠绕组 simplex lap winding
单独试验 individual test
单发动机 monomotor
单缸汽轮机 single-cylinder machine
单个装配 individual mounting
单环路系统 one-loop system
单晶(体) monocrystal
单块 monoblock
单脉冲 monopulse
单模光纤 monomode fibre
单片放大器 one-chip amplifier
单片混合电路 monobrid circuit
单片组装法 monobrid
单位长度电感量 inductance per unit length
单位电机体积功率 horsepower per machine volume
单位制 system of units
单稳态继电器 monostable relay
单相电路 single-phase circuit
单相系统 single-phase system
单像管 monoscope
单液电池 one-fluid cell
单音脉冲发生器 tone-burst generator
单元 element
单元机组 monoblock
单元接线图 unit connection diagram
单匝感应器 loop inductor
单轴汽轮机 tandem turbine
单轴双排汽轮机 tandem double-flow turbine
单轴(系)汽轮机 tandem compound turbine
氮 nitrogen
挡板 baffle
挡块 stop
挡水板 water baffle
档距 span
刀柄 shank
刀开关 knife switch
刀形触头 knife contact
刀闸阀 KGV (knife gate valve)
导波体 wave director
导电性 conductibility; conductivity
导航信号发送机 transmityper
导缆器 fairlead
导索器 fairlead
导通电阻 on-resistance
导线 lead
导线点 traverse point
导向力 guidance force
导向轮 idler
岛 island
倒电容 elastance
倒反 inversion
倒钩 beard
倒虹吸管 inverted siphon
倒介电常数 elastivity
倒数 reciprocal
倒转 inversion
道岔 track switch
道密度 track density
灯标 beacon
灯管 lamp tube
灯具 lighting fitting
灯具污垢减光 luminaire dirt depreciation
灯泡钨丝 osram
灯丝 filament
灯头 base
灯芯 lamp mount

灯罩 shade
灯座 socket
登记 logging
等待时间 latency (time)
等发光强度线 isocandela diagram
等分回转工作台 indexing table
等高距 contour interval
等高线 contour
等级 rating
等坎德拉图 isocandela diagram
等雷电级 isokeraunic level
等离子弧 plasma arc
等离子喷涂 plasma spraying
等离子体 plasma
等离子体频率 Langmuir frequency
等力调节器 isodynamic governor
等熵指数 isentropic exponent
等声强曲线 isoacoustic curve
等时调节器 isochronous governor
等温控制器 isothermal controller
等压 constant pressure
低低 LL (low-low)
低电平逻辑电路 low-level logic (circuit)
低功率传输 low-level transmission
低能电子衍射 LEED (low-energy electron diffraction)
低频 LF (low frequency)
低频扼流圈 LFC (low-frequency choke)
低频滤波器 LFF (low-frequency filter)
低频噪音 hum
低损耗光纤 low-loss fibre
低温冷却 subcooling
低温学 cryogenics
低压 LP (low pressure)
低压熔断器 low-voltage fuse
低阻抗传输 LIT (low-impedance transmission)

滴定管 burette
滴浸树脂 trickle resin
底板 bedplate
底流 underflow
底漆 priming
底视图 backplan
地 earth
地层 stratum
地磁 earth magnetism
地磁场 earth magnetic field
地磁感应 earth induction
地磁感应器 earth inductor
地电势 earth potential
地电位 earth potential
地电位差 earth potential difference
地电位作业 earth potential working
地回电路 earth return circuit
地连接线 ground conductor
地貌 topographic feature
地面电缆 land cable
地面耦合干扰 ground-coupled interference
地面信号板 ground signal panel
地下电缆 underground cable
地下排水系统 subdrainage
地线 earth wire; ground (wire)
地线夹 ground clamp
递变电阻分压器 tapered potentiometer
第三绕组 tertiary winding
第一基点 first base point
第一临界转速 first critical speed of rotation
典型栅元 typical cell
点到点控制系统 point-to-point control system
点定位 point location
点动 inching
点动模式 inching mode
点动速度 inching speed
点对点传输 point-to-point transmission
点对点控制 point-to-point control

点对点连接 point-to-point connection
点分辨力 point-to-point resolution
点焊 spot welding
点火电流 ignition current
点火电路 ignition circuit
点火管 ignitron
点火极 igniter; ignitor
点火剂 igniter; ignitor
点火器 igniter; ignitor
点火相互作用 ignition interaction
点火油枪 torch oil gun
点接触 point contact
点模式转换 inversion of point patterns
点漂 point drift
点蚀斑 pitting
点头灯丝 point filament
点位控制 point-to-point control
碘化钠 sodium iodide
电表键 meter key
电波发射器 wave launcher
电波辐射 wave radiation
电波吸收体 wave absorber
电波仰角 wave angle
电波遥测(术) wave telemetry
电测程 electrologging
电测高温计 electropyrometer
电测角器 electrogoniometer
电测角仪 electrogoniometer
电测力计 electrodynamometer
电测湿度计 electropsychrometer
电测温度计 electrothermometer
电测针 electroprobe
电厂 power station
电厂配套设施 BOP (balance of plant)
电场 electric field
电场强度 electric field intensity
电场线 electric field line
电沉积 electrodeposition
电磁 electromagnetism
电磁场 EMF (electromagnetic field)
电磁成形 electromagnetic forming
电磁放大透镜 EAL (electromagnetic amplifying lens)
电磁辐射 EMR (electromagnetic radiation)
电磁干扰 EMI (electromagnetic interference)
电磁感应 electromagnetic induction
电磁感应定律 Faraday's law (of induction); law of electromagnetic induction
电磁继电器 electromagnetic relay
电磁兼容性 EMC (electromagnetic compatibility)
电磁接触器 electromagnetic contactor
电磁量 EMV (electromagnetic volume)
电磁体 electromagnet
电磁铁 electromagnet
电磁线圈 electromagnetic coil
电磁学 electromagnetics; electromagnetism
电磁转矩 electromagnetic torque
电刀 electrotome
电导 conductance
电导计 conductometer
电导率 (electric) conductivity
电导仪 conductometer
电动泵 motor pump
电动测功机 electrical dynamometer
电动机 dynamo
电动机式仪表 motor meter
电动力 electrodynamic force
电动力学 electrodynamics; electrokinetics
电动式继电器 electrodynamic relay
电动势 electrodynamic potential; electromotance; EMF (electromotive force)
电动天平 electrobalance

电动钻具 electrodrill
电度 electrical degree
电度表 watt-hour meter
电镀 electrofacing; electroplating; galvanization; galvanoplasty
电镀金 electrogilding
电镀物 electroplate
电锻 electro-forging
电反射比 electroreflectance
电反射率 electroreflectance
电分解作用 electrode composition
电分散作用 electrodispersion
电腐蚀 electrocorrosion
电负性气体 electronegative gas
电感 inductance
电感表 inductance meter
电感电容耦合 inductance-capacitance coupling
电感器 inductor
电感式位移传感器 inductive displacement transducer
电感性负载调节 inductive load adjustment
电感应 electric induction
电工 electrotechnics
电工铝 high-conductivity aluminium
电工学 electrotechnology
电共振 electric resonance
电光学 electrooptics
电焊 electric soldering; electrowelding
电合成 electrosynthesis
电荷 electric charge
电痕 track
电弧 arc
电弧发生器 igniter; ignitor
电弧焊 arc welding
电弧能 arc energy
电弧熔化 arc melting
电化当量 electro-equivalent
电化学放电加工 ECDM (electrochemical discharge machining)

电火花点火器 electric spark igniter
电火花加工 EDM (electro-discharge machining)
电击 electric shock
电机械研究 EMR (electromechanical research)
电极 electrode
电极壁 electrode wall
电极臂间距 throat gap
电极臂伸出长度 throat depth
电极电导 electrode conductance
电极电容 electrode capacitance
电极电势 electrode potential
电极电位 electrode potential
电极端 electrode edge
电极间隙 electrode gap
电极偏压 electrode bias
电极台板 platen
电极握臂 electrode arm
电集流 electrojet
电加工 electromachining
电键 key
电键频率 keying frequency
电接触 electric contact
电接触不良 intermittent electric contact
电解 electrolysis
电解槽 electrolysis tank; electrolyzer
电解沉积 electrolytic deposition
电解池 electrolytic bath; electrolyzer
电解萃取 electroextraction
电解电容器 electrolytic capacitor
电解电阻 electrolytic resistance
电解分析 electrolytic analysis
电解精炼 electrorefining
电解器 electrolyzer
电解溶解 electrodissolution
电解提取 electrowinning
电解液 electrolyte
电解质 electrolyte
电解装置 electrolyzer
电介质 dielectric

电精制　electrorefining
电抗部分　wattless component
电抗电桥　X-bridge
电抗电压　reactance voltage
电抗继电器　reactance relay
电抗器　reactor
电抗线圈　reactance coil
电刻　electrogravure
电空制动器　electropneumatic brake
电控重力　electrogravity
电缆套管　joint box
电离　ionization
电离比　ionization ratio
电离电压　ionization voltage
电离放电　ionization discharge
电离辐射　ionizing radiation
电离概率　ionization probability
电离管　ionization tube
电离起始电压　ionization inception voltage
电离探测器　ionization detector
电离熄火电压　ionization extinction voltage
电离现象　ionization phenomenon
电离阈值　ionization threshold
电离杂质　ionizing impurity
电力测功法　electrodynamometry
电力测功计　electrodynamometer
电力电缆　power cable
电力电源　EPS (electric power supply)
电力交易　PE (power exchange)
电力推进　EP (electric propulsion)
电力系统自动化　power system automation
电力线　electric field line
电量计　electricity meter
电零(点)　EZ (electrical zero)
电零位调节器　electrical zero adjuster
电流　current
电流变送器　CT (current transmitter)
电流表　ampere-meter
电流电压　galvanic voltage
电流互感器　CT (current transformer)
电流恢复比　current recovery ratio
电流继电器　current relay
电流密集区　high-current-density region
电流耦合　galvanic coupling
电流误差　current error
电流引入　current injection
电流元　current element
电炉变压器　furnace transformer
电炉渣　electroslag
电路　circuit
电路内部仿真器　in-circuit emulator
电路耦合　hook-up
电纳继电器　susceptance relay
电黏滞性　electroviscosity
电偶极子　electric dipole
电抛光　electropolishing
电喷镀金属　electrometallization
电喷流　electrojet
电平表　LM (level meter)
电平触发器　LT (level trigger)
电平计　LM (level meter)
电平控制　levecon (level control)
电平漂移二极管　level-shifting diode
电平漂移放大器　LSA (level-shifting amplifier)
电平调节　level adjustment
电平图　level diagram
电屏蔽　electric shielding
电气　electricity
电气传导　electric conduction
电气发射器　electropult
电气气动接触器　electropneumatic contactor
电气阻尼器　electrical damper
电热材料　thermo-electric material
电热的　electrothermal

电热学 electrothermics; electrothermy
电容 capacitance
电容器 capacitor; condenser
电容器〖带熔断器的〗 fused capacitor
电容式电压互感器 CVT (capacitor voltage transformer)
电容效应 effect of capacitance
电溶胶 electrosol
电渗 electroosmosis
电渗析 electrodialysis
电生理学 electrophysiology
电湿度计 electropsychrometer
电湿法冶金 electro-hydrometallurgy
电蚀刻 electroetching; electrograving
电枢 armature
电刷 brush
电刷内电压降 internal brush drop
电缩作用 electrostriction
电梯用电缆 travelling cable
电通(量) electric flux
电通密度 electric flux density
电透析 electrodialysis
电推进 EP (electric propulsion)
电网供电 mains supply
电位差 (electric) potential difference
电位差计 potentiometer
电位差计残余电动势 residual electromotive force of potentiometer
电位(滴定)法 potentiometry
电位记录器 electrograph
电位器 potentiometer
电位器罗盘 potentiometer compass
电位器式传感器 potentiometric sensor; potentiometric transducer
电位器式位移传感器 potentiometric displacement transducer
电位器式压力传感器 potentiometer (type) pressure transducer
电位式分析器 potentiometric analyzer
电物理学 electrophysics
电吸附 electrosorption
电析 electroextraction
电线 cord
电线管 conduit
电响应 electroresponse
电信号机 electrosemaphore
电压 electric tension
电压比 voltage ratio
电压比〖电容器式分压器的〗 voltage ratio of a capacitor divider
电压表 voltmeter
电压波动 voltage fluctuation
电压常数 voltage constant
电压传感器 voltage sensor; voltage transducer
电压放大器 voltage amplifier
电压负反馈 negative voltage feedback
电压互感器 voltage transformer
电压回路 voltage circuit
电压尖脉冲缓冲电路 snubber
电压降 voltage drop
电压控制门组件 voltage controlled gate module
电压灵敏度 voltage sensitivity
电压敏感器 voltage sensor
电压敏感性 voltage sensitivity
电压匹配互感器 voltage matching transformer
电压调整(率) voltage regulation
电压误差 voltage error
电压中心 voltage centre
电压骤降陡度 steepness of voltage collapse
电冶金法 electrometallurgy
电冶金学 electrometallurgy
电液成形 electrohydraulic forming
电液控制 EHC (electrohydraulic control)

电液执行机构 electrohydraulic actuator
电泳 electrophoresis
电泳图 electrophoretogram
电源 power source
电源电压 mains voltage; power supply voltage
电源焊接电压 power source welding voltage
电源开关 mains switch
电源频率 mains frequency; power supply frequency
电源装置 power supply device
电晕 corona
电造石英 electroquartz
电渣 electroslag
电渣焊机 electroslag welder
电站厂房 powerhouse
电振动器 electrovibrator
电致发光 EL (electroluminescence)
电致伸缩 electrostriction
电滞 electric hysteresis
电中微子 electrino
电铸 electroforming; electrotyping; galvanoplasty
电铸术 galvanoplasty
电子 electron
电子测微法 electromicrometry
电子测微计 electromicrometer
电子传导 electron conduction
电子导电 electron conduction
电子导纳 electron admittance
电子电导率 electronic conductivity
电子电离 electron ionization
电子电路分析程序 ECAP (electronic circuit analysis program)
电子对 electron pair
电子对抗 ECM (electronic countermeasures)
电子仿生学 electronic bionics
电子俘获 electron capture
电子干扰 ECM (electronic countermeasures)

电子管 electron tube; radio tube
电子光学 electron optics
电子加速器 electron accelerator
电子空位 electron vacancy
电子耦合 electron coupling
电子耦合控制 ECC (electron coupling control)
电子漂移 electron drift
电子亲和势 electron affinity
电子束 e-beam; electron beam
电子束曝光系统 electron beam exposure system
电子束成像 electron beam patterning
电子束发生器 electron beam generator
电子束记录仪 EBR (electron beam recorder)
电子束扫描系统 EBS system (electron beam scanning system)
电子稳速器 electronic governor
电子衍射 electron diffraction
电子异构 electromerism
电子异构体 electromer
电子异构物 electromer
电子异构作用 electromerization
电子跃迁 electron transition
电子振荡器 electronic generator
电子制导 EG (electronic guidance)
电子自动交换机 EAX (electronic automatic exchange)
电阻 resistance; resistor
电阻表 ohmmeter
电阻电桥 ohmic bridge
电阻电压降 IR drop; ohmic drop
电阻干湿球湿度计 resistance psychrometer
电阻焊 ERW (electrical resistance welding)
电阻率 resistivity
电阻平衡 resistance balance

电阻器 resistor
电阻湿度计 resistance hygrometer
电阻式测斜仪 resistance inclinometer
电阻损失 ohmic loss
电阻箱 resistance box
电阻箱残余电感 residual inductance of resistance box
电阻箱残余电阻 residual resistance of resistance box
垫块 bolster
垫片 filler strip; shim
垫圈 gasket
垫条 filler strip
吊车 crab
吊灯 pendant (fitting)
吊杆 suspender
吊杆柱 king post
吊索 suspender
调度 scheduling
调用 invocation
迭代因子 iteration factor
迭代语句 iteration statement
叠层天线 laminated antenna
叠片 lamination
叠片磁铁 laminated magnet
叠片绝缘 lamination insulation
叠片漆 lamination varnish
叠片铁芯 laminated core
叠绕组 lap winding
蝶阀 BFV (butterfly valve)
蝶式绝缘子 shackle insulator
蝶形阀 BFV (butterfly valve)
丁烷 butane
顶板 top slab
顶部悬吊支承锅炉 top-supported boiler
顶出器 knockout
顶轭 top yoke
顶面标高 top elevation
顶脱件 knockout
顶圆直径 tip diameter
顶置喷燃器 top-fired burner
顶轴油泵 jacking oil pump
定常调整器 time-invariant regulator
定常系统 time-invariant system
定额 quota; rating
定量差热分析 QDTA (quantitative differential thermal analysis)
定量差热分析仪 quantitative differential thermal analyzer
定量化 quantification
定时波 timing wave
定时磁电机 timed magneto; timing magneto
定时电动机 timing motor
定时电路 timing circuit
定时电压 timing voltage
定时断流器 time cut-off
定时断路器 time cut-out
定时继电器 timed relay; time relay
定时接触器 timing contactor
定时开关 time switch
定时脉冲 timed pulse; timing pulse
定时脉冲继电器 time pulse relay
定时器 keyer; timer
定位焊 tack welding
定位环 locating ring
定位件 keeper
定位螺栓 jack bolt
定位器 localizer; locator; positioner
定位式按钮 locked push button
定位销 knock pin
定位装置 locator
定温开关 thermostatic switch
定限继电器 marginal relay
定向结晶 oriented crystallization
定性分析 qualitative analysis
定性物理模型 qualitative physical model
定压 constant pressure
定压启动 constant pressure start-up
定制电动机 tailor-made motor

定子 stator
定子绕嵌机 stator winding machine
锭子 spindle
氡 radon
动臂起重机 derrick
动电计 electrokinetograph
动电学 electrokinetics
动合触点 make contact
动合触头 make contact
动合输出电路 output make circuit
动静万能试验机 static/dynamic universal testing machine
动理学 kinetics
动力电流 kinetic current
动力平衡 kinetic equilibrium
动力学 dynamics; kinetics
动量 (kinetic) momentum
动能 kinetic energy
动态负荷 dynamic load
动态平衡 kinetic equilibrium
动态稳定 dynamic stability
动作泄漏电流 leakage operating current
冻结 congelation
冻凝 congelation
冻土学 cryopedology
抖动 jitter
独立菜单 isolated menu
独立电厂 isolated power plant
独立电流源 independent current source
独立循环电路组件 independent circulating circuit component
独立循环回路部件 independent circulating circuit component
独立镇流器 independent ballast
读出电路 sense circuit
堵转电流 locked-rotor current
堵转试验 locked-rotor test
度 kWh (kilowatt-hour)
渡越时间管 transit-time tube
镀层 facing
镀锡 tinning
镀锌 galvanization; zincification
镀锌试验 galvanizing test
端部负载渐减电缆 taper-loaded cable
端差 TD (terminal difference)
端盖 end bracket
端接错误 mistermination
端接式电阻箱 resistance box with terminals
端面齿距 transverse pitch
端头配件 end fitting
端线 terminal wire
端子板 TB (terminal board)
端子排 TB (terminal board)
短截线 stub
短路 short circuit
短路保护变压器 short-circuit-proof transformer
短路比 short-circuit ratio
短路电流 fault current
短路换向器试验 short-circuit commutator test
短路接通 short-circuit making
短路容量 short-circuit capacity
短路线圈测试仪 growler
短路运行 short-circuit operation
短路匝 short-circuit turn
短路阻抗 fault impedance
短时运行功率 time rating
短暂故障 transient fault
短轴 minor axis
段间绝缘 section insulation
断电 de-energize; interruption of power supply
断电时间 interruption duration
断开 interruption; opening
断开电流 interrupting current
断开电路 OC (open circuit)
断开容量 interrupting capacity
断开状态 off state
断裂韧性 fracture toughness
断流阀 stop valve
断流下限额定值 minimum current interrupting rating
断路电键 interruption key
断路电流 interruption current
断路媒质 interrupting medium

断路器 circuit-breaker; interrupter; killer
断路器〖带熔断器的〗 integrally fused circuit-breaker
断路器电容器 circuit-breaker capacitor
断路容量 interrupting capacity
断路时间 interrupting time
断路试验 interrupting test
断路装置 interrupting device
断面收缩率 reduction of area
断水器 water circuit-breaker
断态 off state
断态电流 off-state current
断态电压 off-state voltage
断线 killed line
断相继电器 open-phase relay
断相运行 open-phase running
断续电流 intermittent current
断续额定值 interrupting rating
断续工作制 intermittent duty
断续故障 intermittent fault
断续进给 intermittent feed
断续流通 intermittent flow
断续试验 interrupting test
断续输入 interrupt input
断续运行试验 intermittent service test
断续装置 make-and-break device; ticker
堆叠器 stacker
堆料机 stacker
堆取料机 stacker-reclaimer
堆芯 core
队列控制块 QCB (queue control block)
对比试块 reference block
对策论 game theory
对称短路电流初始值 initial symmetrical short-circuit current
对称性 symmetry
对冲燃烧 opposed firing
对地电容 earth capacitance
对地短路电流 ground-fault current

对地短路探测 ground leakage detection
对地高电阻故障 high-resistance fault to earth
对地高阻抗故障 high-impedance fault to ground
对地净空距离 ground clearance
对地绝缘 earth insulation; ground insulation; isolation from earth
对地漏电故障 earth leakage fault
对地泄漏 earth leakage
对接触头 butt contact
对接焊缝 BW (butt weld)
对接接触 butt contact
对流 convection
对流器 convector
对数变换伏特计 LCVM (log conversion voltmeter)
对消法 opposition method
钝化 passivation
盾构法 shield method
多标度仪表 multi-scale instrument
多波段扫描微波辐射仪 SMMR (scanning microwave multiband radiometer)
多波束 multibeam
多参数调整 multiparameter regulation
多层 multilayer
多层基片 multilayer substrate
多层掩膜 layered mask
多层阴极 multilayer cathode
多重编程器 gang programmer
多重处理 multiprocessing
多重处理系统 multiprocessing system
多重连接 multiple connection
多重频率 multiple frequency
多重矢量 multivector
多重图像 multi-image
多抽头绕组 tapped winding
多抽头线路 tapped line
多传感头 multihead

多道处理 multiprocessing
多点取样 multidraw
多电荷离子 multiply charged ion
多电极 multiple electrode
多电平编码 multilevel encoding
多断开关 multibreak switch
多发射极晶体管 multi-emitter transistor
多分支电缆 multiway cable
多功能测量仪表 multi-function measurement instrument
多功能电能仪表 combimeter
多功能仪表 multi-purpose instrument
多光束 multibeam
多光子过程 multiphoton process
多光子吸收 MPA (multiphoton absorption)
多环的 multiloop
多回路的 multiloop
多回路控制系统 multiloop control system
多级 multistage
多级放大器 multi-amplifier; multistage amplifier
多级平行弧焊 tangent welding
多极 multiple electrode
多极电机 multipolar machine
多极熔断器 multipole fuse
多脚架 spider
多接头线圈 tapped coil
多节点 multinode
多节滤波器 multisection filter
多孔磁芯存储器 transfluxor
多孔隔板阴极 L-cathode
多框铁芯 multiframe core
多联开关 multigang switch
多量程电压表 multivoltmeter
多量程伏特计 multivoltmeter
多量程仪表 multi-range instrument
多路传输 multiplex
多路的 multiway
多路定标器 multiscaler
多路分配器 demultiplexer
多路复用 multiplex
多路开关 multiway switch
多路耦合器 multi-coupler
多路绕组 multicircuit winding
多模光纤 multimode fibre
多频带 multiband
多频系统 multi-frequency system
多腔滤波器 multicavity filter
多区炉 multi-zone furnace
多绕组变压器 multi-winding transformer
多栅管 multigrid tube
多栅极 multigrid
多数决定门 majority gate
多数载流子 majority carrier
多速电动机 multi-speed motor
多通道蠕动泵 multi-channel peristaltic pump
多筒式烟囱 multitube stack
多头感应线圈 tapped induction coil
多头线圈 tapped coil
多位置的 multiposition
多位置元件 multiposition element
多线圈的 multi-turn
多线圈电位器 multi-turn potentiometer
多向 multiway
多向开关 multiway switch
多相 multiphase
多相重合闸 multiphase reclosing
多谐振荡器 multivibrator
多芯电缆 multicore cable
多引线的 multilead
多引线装置 multi-outlet assembly
多用表 multimeter; multitester
多用示波器 multi-purpose oscilloscope
多元件绝缘子 multi-element insulator
多匝的 multiloop; multi-turn
多值非线性 multivalued nonlinearity

多址　multiple access
垛　stack
惰轮　idler
惰性气体　inert gas

惰性气体保护焊　inert-gas-shielded welding
惰性气体密封　inert-gas seal

Ee

锇 osmium
锇钨灯丝合金 osram
额定变比 rated transformation ratio
额定出力 rated output
额定电流 nominal current
额定电压 nominal voltage
额定电阻 nominal resistance
额定动态电流 rated dynamic current
额定短时热电流 rated short-time thermal current
额定二次电流 rated secondary current
额定二次电压 rated secondary voltage
额定范围 rated range
额定负载 rated burden; rated load
额定工况 rated operating condition
额定工作范围 rated operating range
额定工作规程 rated operating specification
额定功率 rating
额定静态横向负荷 rated static transverse load
额定静态纵向负荷 rated static longitudinal load
额定连续热电流 rated continuous thermal current
额定流量系数 rated flow coefficient
额定容量 nominal capacity; rated capacity
额定时间 time rating
额定使用范围 rated range of use
额定输出 rated output
额定瓦数 wattage rating
额定一次电流 rated primary current
额定一次电压 rated primary voltage
额定载荷 rated load
额定正弦激振力 rated sine excitation force
额定值 rating
扼流圈 inductor
轭安匝 yoke ampere-turns
轭磁通 yoke flux
轭空气隙 yoke air gap
二次被覆层 secondary coating
二次电流 secondary current
二次离子质谱学 SIMS (secondary ion mass spectroscopy)
二次绕组 secondary winding
二端电路 two-terminal circuit
二端对网络 two-port network
二端口网络 two-port network
二极管 diode
二极管阀 diode valve
二极管-晶体管逻辑 DTL (diode-transistor logic)
二极真空整流管 kenotron
二进码十进数 BCD (binary-coded decimal)
二进制 binary
二进制位 bit
二相发电机 diphaser
二相理论 two-phase theory
二相流 two-phase flow
二氧化硅 silica

二氧化钛 titanium dioxide
二氧化物 dioxide
二自由度 TDF (two degrees of freedom)
二自由度陀螺仪 two-degree-of-freedom gyroscope

Ff

发电厂 generating plant; generating station
发电机 dynamo; generator
发电机变压器 generator transformer
发电机端子 generator terminal
发电机额定值 generator rating
发电机功率定额 generator rating
发电机机坑 generator pit
发电机母线 generator bus
发电机输出 generator output
发电机引线 generator bus; generator lead
发电机转子 generator rotor
发电机组 generating set
发电能力 generating capacity
发电容量 generating capacity
发电设备 generating set
发电站 powerhouse
发光二极管 LED (light-emitting diode)
发光开关 illuminated switch
发光模拟图 illuminated mimic diagram
发起站 initiator
发散 divergence
发散冷却叶片 transpiration-cooled blade
发射 emission
发射电流 emission current
发射换能器 transmitting transducer
发射机分配器 TD (transmitter distributor)
发射频率 emission frequency
发生器 generator

发送磁芯 transmitting core
发送晶体 transmitting crystal
发送元件 transmitting element
乏表 varmeter
乏时计 var-hour meter
阀 valve
阀闭锁 valve blocking
阀尺寸计算 valve sizing
阀出口马赫数 valve outlet
阀电抗器 valve reactor
阀杆 valve stem
阀杆不平衡 valve stem unbalance
阀基 valve base
阀口径计算 valve sizing
阀片 valve element
阀容量 valve capacity
阀体 valve body
阀芯 valve plug
阀轴 valve shaft
阀座 valve seat
法 farad
法定安培 legal ampere
法定伏特 LV (legal volt)
法定欧姆 legal ohm
法拉 farad
法拉第定律 Faraday's law (of induction)
法线轴 normal axis
法向导数 normal derivative
翻料架 tilting trestle
翻转炉排片 tilting bar
反冲 kick; kickback
反冲力 kick
反冲式起动器 kick starter
反动式汽轮机 reaction turbine
反虹吸 back siphonage
反接电动机 inverted motor

反接法 opposition method
反馈 feedback
反馈环节 feedback element
反馈控制 feedback control
反馈线圈 tickler (coil)
反力涡轮 reaction turbine
反射系数 reflection coefficient
反射照明 incident illumination
反时限过(电)流保护 inverse-time overcurrent protection
反时延动作 inverse time-delay operation
反时延过(电)流脱扣器 inverse time-delay overcurrent release
反速电动机 inverse-speed motor
反弹 bounce
反向 inversion
反向变换 inverse transformation
反向并联 inverse parallel connection
反向磁化的 inversely magnetized
反向电极电流 inverse electrode current
反向电抗 inverse reactance
反向电流 inverse current
反向负阻抗变换器 inverting negative impedance converter
反向激励 inverse excitation; negative excitation
反向励磁 inverse excitation; negative excitation
反向偏压 reverse bias
反向旋转变流机 inverted rotary converter
反向阻断二极晶闸管 reverse blocking diode thyristor
反相 opposite phase
反相放大器 inverting amplifier
反相器 inverter; negater
反型层 inversion layer
反絮凝 deflocculation
反演 inversion
反演密度 inversion density
反应 reaction
反应堆 reactor

反转电机 inverted machine
反转控制器 inverse-acting controller
反作用力 reaction
返程 kickback
范围 orbit
方波冲击电流 rectangular impulse current
方块图 block diagram
方铁矿 wustite
方头螺栓 lag bolt
方位 bearing
方向继电器 directional relay
防爆率 knock rating
防擦物 fender
防超温保护 overheating protection
防滴式 drip-proof type
防腐剂 aseptic
防垢处理 scale prevention treatment
防护频带 guard band
防护屏 shield
防护式加热元件 protected heating element
防火花的 spark-proof
防漏的 leakproof
防喷型电机 hose-proof machine
防渗设施 seepage control installation
防水层 waterproof course; waterproof layer
防水的 waterproof
防水粉 waterproofing powder
防水机械 impervious machine
防水剂 waterproofing agent
防水浆 water-resistant solution
防水结构 waterproof structure; watertight construction
防水绝缘 waterproof insulation
防水砂浆 waterproof mortar
防水外壳 waterproof case
防水型电动机 waterproof motor
防水型电机 waterproof machine
防水型电气设备 waterproof electrical equipment

防水型感应电动机 waterproof induction motor
防水型仪表 waterproof instrument
防水毡 waterproof carpet
防锈包装 rust-preventive packaging; rustproof packaging
防震包装 shockproof packaging
房屋引入电缆 incoming service cable
房屋引入架空电缆 incoming service aerial cable
仿真 simulation
仿真程序 simulation program
仿真方法学 simulation methodology
仿真方框图 simulation block diagram
仿真工作站 simulation work station
仿真过程 simulation process
仿真过程时间 simulation process time
仿真环境 simulation environment
仿真技术 simulation technique
仿真结果 simulation result
仿真框图 simulation block diagram
仿真类型 simulation type
仿真模型 simulation model
仿真模型库 simulation model library
仿真评价 simulation evaluation
仿真器 simulator
仿真软件 simulation software
仿真设备 simulation equipment
仿真时钟 simulation clock
仿真实验 simulation experiment
仿真实验模式库 simulation experiment mode library
仿真实验室 simulation laboratory
仿真数据 simulated data
仿真数据库 simulation database
仿真速度 simulation velocity
仿真算法 simulation algorithm
仿真算法库 simulation algorithm library
仿真图形库 simulation graphic library
仿真系统 simulation system
仿真信息库 simulation information library
仿真语言 simulation language
仿真运行 simulation run
仿真支持系统 simulation support system
仿真知识库 simulation knowledge base
仿真中断 simulated interrupt
仿真中心 simulation centre
仿真专家系统 simulation expert system
仿真作业 simulation job
放大 amplification; magnification
放大镜 magnifier
放大率 magnification
放大器 amplifier
放电 discharge
放电电流 discharging current
放电电渗处理 EDP (electric diffusing process)
放电加工 EDM (electro-discharge machining)
放电器 discharger
放能度 exoergicity
放能反应 exoergic reaction
放气 blow-off
放散管 bleeder
放射色谱法 radio chromatography
放射性测量 radioactive survey
放射性污染 radioactive contamination
放水道 tailrace
放水管 adjutage
飞轮二极管 flywheel diode
非 negation
非层流粒子束 non-laminar beam
非导体 non-conductor

非电解质 non-electrolyte
非电路 NOT circuit
非定界符 non-delimiter
非法操作 illegal operation
非关联失效 non-relevant failure
非极化继电器 non-polarized relay
非极性电介质 non-polar dielectrics
非计划降额 UD (unplanned derating)
非计划降额小时 UDH (unplanned derating hours)
非均匀场 non-uniform field
非连通网络 unconnected network
非门 negater; negation gate; NOT gate
非欧姆律电阻(器) non-ohmic resistor
非偏振光 non-polarized light
非全相运行 open-phase running
非确定性故障 indeterminate fault
非时变系统 time-invariant system
非锁定电键 non-locking key
非锁定继电器 non-locking relay
非弹性碰撞 inelastic impact
非调制 non-modulation
非线性 non-linearity
非线性失真 klirr
非线性调制 non-linear modulation
非线性系数 non-linear factor
非线性相关 non-linear dependence
非相干辐射 non-coherent radiation
非相干探测 non-coherent detection
非旋场 lamellar field
非振荡放电 non-oscillating discharge
非正交性 non-orthogonality
废铬液 waste chrome liquor
废化学试剂 waste chemical reagent
废活性污泥 waste activated sludge
废碱 waste alkali
废碱液 waste lye
废金属 scrap (metal)
废料 waste material; waste stock
废煤 waste carbon
废凝结水水泵 waste condensate pump
废气 waste air
废气处理 waste gas treatment
废气处置系统 waste gas disposal system
废气分析 waste gas analysis
废气过热器 waste gas superheater
废气净化 waste gas cleaning
废气净化设备 waste gas cleaning plant
废气净化系统 waste gas cleaning system
废气排放标准 waste gas emission standard
废气燃烧 waste gas burning
废气脱硫 waste gas desulphurization
废气污染防治 waste gas pollution control
废气系统 waste gas system
废气压缩机 waste gas compressor
废气增压器 waste gas supercharger
废气蒸气捕集器 waste gas vapour trap
废氰化物 waste cyanide
废燃料 waste fuel
废热过热器 waste heat superheater
废热回收 waste heat recovery
废热利用 waste heat utilization
废热排出系统 waste heat removal system

废热损失 waste heat loss
废热烟道 waste heat flue
废砂石 waste sand and gravel
废水 effluent
废水池 waste disposal basin
废水处理 wastewater processing; wastewater treatment
废水处理厂 wastewater treatment plant
废水处理过程 wastewater treatment process
废水处理设施 wastewater facility
废水处理系统 wastewater treatment system
废水处置设施 wastewater disposal facility
废水复用 wastewater reuse
废水固渣 wastewater solid
废水管 wastewater pipe
废水回收 wastewater reclamation
废水回收厂 wastewater reclamation plant
废水净化 wastewater purification
废水净化厂 wastewater purification plant
废水流量 wastewater flow
废水率 wastewater rate
废水收集 wastewater collection
废水收集系统 wastewater collection system
废水稳定塘 waste stabilization pond
废水污泥 wastewater sludge
废水稀释 wastewater dilution
废水消毒剂 wastewater disinfectant
废水氧化池 waste oxidation basin
废水再用 wastewater reuse
废水中和器 wastewater neutralizer
废水中和箱 wastewater neutralization tank
废水组成 wastewater composition
废丝 waste silk
废酸 waste acid
废碳 waste coal
废污泥 waste sludge
废物 waste material
废物安置 waste placement
废物包装 waste package
废物处理 waste treatment
废物处理技术 waste treatment technique
废物处理设备 waste treatment equipment
废物处理设施 waste plant
废物处理站 waste treatment station
废物处置 waste disposal
废物处置库 waste repository
废物堆 waste bank
废物焚烧装置 waste calcining facility
废物粉碎机 waste shredder
废物封存 waste containment
废物封隔 waste confinement
废物负载 waste load
废物固化 waste solidification
废物固化工厂 waste immobilization plant
废物管理 waste management
废物锅炉 waste boiler
废物海床下处置 waste disposal beneath the ocean floor
废物核嬗变处置 waste disposal by nuclear transmutation
废物回收发电厂 waste recovery power plant
废物监测 waste monitoring
废物控制 waste control
废物埋葬 waste burial
废物埋葬场 waste burial ground
废物取样箱 waste sampling cabinet
废物容器 waste container
废物收回系统 waste retrieval system

废物收集系统 waste collector system
废物衰变箱 waste decay tank
废物形态 waste form
废物形态调整 waste conditioning
废物种类 waste category
废物贮存 waste storage
废物贮存场 waste-storage farm
废物贮存区 waste-storage area
废物贮存系统 waste retention system
废物转换技术 waste convertion technique
废物最终处置 ultimate waste disposal
废液 waste fluid
废液除盐器 waste demineralizer
废液除盐装置 waste demineralizing plant
废液返回循环 waste back-cycling
废液回收 waste liquor recovery
废液燃烧装置 waste fluid burning plant
废液蒸发回流冷凝器 waste evaporative reflux condenser
废油 waste oil
废油处置 waste oil disposal
废油凝块 waste oil clot
废铀 waste uranium
废原油 waste crude oil
沸石 zeolite
费拉里斯电表 Ferraris meter
费拉里斯电动机 Ferraris motor
分半槽楔 two-piece wedge
分瓣电刷 split brush
分贝 decibel
分辨力 resolution
分辨率 resolution
分波器 wave separator
分布式混合燃烧器 DMB (distributed mixing burner)
分布因数 spread factor
分部件 sub-assembly
分层 layering

分层接线电缆 horizontal floor wiring cable
分层介质 layered medium
分层绕组 layer winding
分层系数 lamination factor
分叉 bifurcation
分带 zonation
分度工作台 indexing table
分度弧 limb
分度盘 scale plate
分段电压试验 piecewise test
分段绝缘 section insulation
分段励磁绕组电动机 tap-field motor
分段气隙电抗器 tapped air-gap reactor
分段送风 zoned air supply
分段器 sectioning device
分断器 interrupter
分断钥匙 interlock deactivating key
分断装置 interlock deactivating means
分光光度滴定法 spectrophotometric titration
分光光度计 spectrophotometer
分级功率 tapped horsepower rating
分级绝缘 non-uniform insulation
分级绝缘绕组 winding with graded insulation
分接 tapping
分接变换操作 tap-change operation
分接参数 tapping quantities
分接电流 tapping current
分接电压 tapping voltage
分接范围 tapping range
分接级 tapping step
分接接触器 tapping contactor
分接开关 tapping switch; tap-selector
分接开关调节器 tapping adjuster
分接量 tapping quantities

分接容量 tapping power
分接头 tap; tapping point
分接头电压 tap voltage
分接头换接 tap-change operation
分接头换接装置 tap changer
分接头控制 tapped control
分接头切换 tap changing
分接头调整器 tap adjuster
分接头位置信息 tap position information
分接头位置指示器 tap position indicator
分接头选择器 tap-selector
分接因数 tapping factor
分解 decomposition
分类器 classifier
分离 segregation
分离触点 isolating contact
分离点 isolated position; isolating point
分离接地 independent earthing
分离器 separator
分离闸刀 isolating blade
分流 splitting
分流器 splitter
分流线圈 transition coil
分路器 splitter
分配 allocation
分配器 allocator
分配箱 distributor box
分区 partition; zonation
分时 time sharing
分束器 beam splitter
分数槽绕组 fractional slot winding
分体积 partial volume
分线盒 junction box
分线开关 tap switch
分相母线 isolated-phase bus
分相屏蔽电缆 radial field cable
分压比 voltage ratio
分压力 partial pressure
分压器 voltage divider
分压器〖具有辅助支路的〗 voltage divider with auxiliary branch circuit

分用器 demultiplexer
分支点 tapping point
分子 molecule; numerator
分组件 sub-assembly
粉砂 silt
粉饰灰泥 stucco
粉碎机 comminutor; kibbler; shredder
粉土 silt
风电机 WTG (wind turbine generator)
风耗 windage loss
风荷载 wind load
风冷电机 fan-cooled machine
风力发电机组 WTG (wind turbine generator)
风轮 rotor
风门 damper
风速计 anemometer
风向标 vane
封闭 containment
封闭母线 enclosed busbar
封口 sealing
封锁 lockout
峰值 peak (value)
峰值电压表 peak voltmeter
峰值负载工作制 peak-load duty
峰值因数 crest factor; peak factor
蜂房式散热器 honeycomb radiator
蜂房式线圈 lattice coil
蜂音器 ticker
缝隙 opening
缝隙损耗 gap loss
敷管炉墙 tube-lined boiler wall
弗莱明右手定则 Fleming's right-hand rule
伏安 VA (volt-ampere)
伏安表 volt-ampere meter
伏安法 voltammetry
伏安时计 volt-ampere-hour meter
伏尔 VOR (VHF omnidirectional range)
伏特表 voltmeter

扶手弯头 knee
氟氯化碳 CFC (chlorofluorocarbon)
浮充电 floating charge
浮垢 scum
浮力 buoyancy
浮栅雪崩注入MOS存储器 FAMOS memory
浮渣 scum
符号电流 marking current
幅度 amplitude
辐射 radiation
辐射表 radiometer
辐射测温法 radiation thermometry
辐射出射度 radiant exitance
辐射传感器 radiation sensor; radiation transducer
辐射高温计 ardometer; radiation pyrometer
辐射功率 radiant power
辐射监测仪 radiation monitor
辐射量 quantity of radiation
辐射能 radiant energy
辐射能量 quantity of radiant energy
辐射平衡表 radiation balance meter
辐射强度 radiant intensity; radiation intensity
辐射热计 bolometer
辐射热流计 radiation heat flowmeter
辐射试验 radiation test
辐射探测器 radiation detector
辐射通量 radiant flux; radiation flux
辐射温度 radiation temperature
辐射温度传感器 radiation temperature sensor; radiation temperature transducer
辐射温度计 radiation thermometer
辐射效率 radiation efficiency
辐射仪 radiometer
辐射元件 radiant element
辐照饱和电流 irradiation saturation current
辐照度 irradiance
俯仰操纵电动机 tilt motor
辅助报告 auxiliary report
辅助电路 auxiliary circuit
辅助回路 auxiliary circuit
辅助控制盘 ACP (auxiliary control panel)
辅助绕组 teaser winding
辅助设备 ancillary equipment
腐蚀的 corrosive
腐蚀性物质 corrosive
负电池 negative battery
负电流 negative current
负电子 negatron
负电子亲和势 NEA (negative electron affinity)
负电子亲和势阴极 NEA cathode
负分接 minus tapping
负荷 burden
负荷开关 on-load switch
负荷系数 loading factor
负辉光灯 negative-glow lamp
负极 cathode; negative electrode
负介子 negative meson
负离子 negative ion
负励磁 negative excitation
负熵 negentropy
负像 negative image
负性 negativity
负序 negative sequence
负压通风 induced draft
负掩模 negative mask
负载电阻(器) loading resistor
负载换相 load commutation
负载脉冲 load pulse
负载能力 load capacity
负载调节 load adjustment
负载系数 load factor
负载线 load line
负载因数 on-load factor
负阻抗管 kallirotron
附加绕组 auxiliary winding
附加设备 optional equipment
附件 accessory

附着 adhesion
附着力 adhesion
复冰厚度 radial thickness of ice
复波绕组 multiple wave winding
复叠绕组 multiplex lap winding
复对四线组 multiple twin quad
复合误差 composite error
复接 multiple connection
复接机键 multiswitch
复接开关 multiswitch
复用器 multiplexer
复用器滤波器 multiplexer filter
复帧 multiframe
副瓣 minor lobe
副励磁机 pilot exciter
傅立叶级数 Fourier series

Gg

伽马分布 gamma distribution
伽马射线照片 gammagraph
伽马射线照相术 gamma-radiography
改进剂 improver
钙 calcium
干板 X 射线照相术 xeroradiography
干度 dryness
干馏 carbonization
干凝胶 xerogel
干扰 interference
干扰场强 interference field strength
干扰电压 interference voltage
干扰电压测量仪 interference voltage meter
干扰滤波器 interference filter
干扰滤片 interference filter
干扰衰落 interference fading
干扰限值 limit of interference
干扰效应 interference effect
干扰抑制 interference suppression
干扰抑制扼流圈 interference suppression choke
干扰抑制符号 interference suppression symbol
干扰抑制装置 interference suppression equipment
干扰源 interference source
干涉 interference
干燥剂 desiccant
干燥无灰基 daf basis (dry ash-free basis)
杆 bar
杆塔 tower
杆体 core
感光乳胶 emulsion
感生通量 induction flux
感应 induction
感应变频机 induction frequency converter
感应场 induction field
感应传感器 inductive pickup
感应点火 induced ignition
感应点火器 induced ignition
感应电流 induced current
感应电位器 inductive potentiometer
感应电压 induced voltage
感应发电机 induction generator
感应分断能力 inductive breaking capacity
感应分流器 inductive shunt
感应过电压耐受试验 induced overvoltage withstand test
感应计数器 induction meter
感应控制电压 induced control voltage
感应雷击 indirect lightning strike
感应炉 induction furnace
感应率 inductivity
感应率效应 effect of inductivity
感应能力 inductivity
感应起电机 influence machine
感应起动器 induction starter
感应圈 induction coil
感应剩余电压 inductive residual voltage
感应式电度表 induction meter
感应线圈 induction coil
感应镇流器 inductive ballast

感应子同步电动机 inductor-type synchronous motor
干线 main line
干线切断 TCO (trunk cut-off)
刚度 rigidity; stiffness
刚性 rigidity; stiffness
刚性联轴器 rigid coupling
刚性转子 rigid rotor
钢 steel
钢带铠装 tape armour
钢化玻璃 toughened glass
钢筋混凝土 reinforced concrete
钢瓶 cylinder
钢丝规 steel wire gauge
钢芯铝线 ACSR cable (aluminium conductor steel-reinforced cable)
杠杆 lever
杠杆开关 lever switch
杠杆式传动 lever drive
杠杆锁 lever lock
高变频机组 high frequency-changer set
高层自动化系统 higher-level automation system
高程 elevation
高纯水 ultrapure water
高磁场超导体 high-field superconductor
高次时延 higher-order time delay
高电导率聚合物 high-conductivity polymer
高电导率铝 high-conductivity aluminium
高电导率铜 high-conductivity copper
高电平电压 high-level voltage
高电平输入电流 high-level input current
高电平信号 high-level signal
高电势 high potential
高电位 high potential
高电压耐用性 high-voltage endurance
高分辨率成像光谱仪 high-resolution image spectrometer
高分断能力熔断器 high-breaking-capacity fuse; high-interrupting-capacity fuse
高分子 macromolecule
高负荷率电价 high-load-factor tariff
高负荷率用户 high-load-factor consumer
高负荷小时 high-load hours
高级语言 high-level language
高阶延迟元件 higher-order delay element
高精度测量仪表 high-precision meter
高灵敏度放大器 high-sensitivity amplifier
高岭石 kaolinite
高岭土 kaolin
高频电容 high-frequency capacitance
高频焊接 high-frequency welding
高频失真试验 high-frequency distortion test
高速磁吹式断路器 high-speed air magnetic breaker
高速反应调压器 high-response-rate voltage regulator
高速继电器 fast relay
高速气吹式断路器 high-speed air-blast breaker
高响应速率调压器 high-response-rate voltage regulator
高效率冲洗器 quick dump rinser
高效率二极管 efficiency diode
高压传感器 high-voltage pickup
高压端子 high-voltage terminal
高压放电灯 high-pressure discharge lamp
高压分压器 high-voltage divider
高压供电 high-voltage power supply
高压接头 high-voltage terminal
高压开关 high-voltage switch
高压配电 high-voltage distribution

高压绕组 high-voltage winding
高压试验 high-voltage test
高压试验技术 high-voltage test technique
高压调压变压器 high-voltage regulating transformer
高压延时 high-voltage delay time
高压真空接触器 high-voltage vacuum contactor
高压整流二极管 kenotron
高压直流 HVDC (high-voltage direct current)
高压直流变电站 HVDC substation
高压直流变压器 HVDC transformer
高压直流换流变压器 HVDC converter transformer
高压直流输电 HVDCT (HVDC transmission)
高烟囱源 tall stack source
高阻表 earthometer; tramegger
高阻抗差动保护 high-impedance differential protection
高阻抗故障 high-impedance fault
锆酸盐 zirconate
锆锡合金 zircaloy
格网 screen
隔板 diaphragm; separator
隔爆型电气设备 flameproof electrical apparatus
隔断刀闸 disconnector
隔行交错显示 interlaced display
隔开物 partition
隔离 insulation; isolation
隔离测量端子 isolating measuring terminal
隔离层 separator
隔离插塞连接器 isolating plug connector
隔离点 isolating point
隔离电感器 isolating inductor
隔离电压 isolation voltage
隔离电压互感器 isolating voltage transformer
隔离度 isolation
隔离端子 isolating terminal
隔离段 isolating distance
隔离放大器 isolation amplifier
隔离开关 disconnector; isolating link; isolating switch; isolator
隔离扩散 isolation diffusion
隔离器 isolator
隔离输出 isolated output
隔离输入 isolated input
隔离位置 isolated position
隔离物 spacer
隔离相母线 isolated-phase bus
隔离中性点端子 isolating neural terminal
隔离子模块 isolating submodule
隔热屏 heat shield
隔热水箱 water tank for heat insulation; water tank for thermal insulation
隔叶块 spacer
隔音板 Celotex board
隔振 vibration isolation
隔振器 vibration isolator
镉 cadmium
各向异性 anisotropy
根据质量选择 QBS (quality-based selection)
跟踪电路 tracking circuit
跟踪滤波器 tracking filter
跟踪锁定 lock-on
跟踪网络 tracking network
跟踪与同步放大器 track-and-hold amplifier
工厂序列号 factory serial number
工频 power frequency
工序间检验 in-process inspection
工业机器人数据接口 industrial robot data interface
工业配电装置 industrial distribution equipment
工艺及仪表流程图 P & ID (process and instrument diagram)

工艺性能试验　processibility test
工质　working medium
工作存储器　work storage
工作电压　operating voltage; WV (working voltage)
工作反向　inverse direction of operation
工作负载　LL (live load)
工作接地　working earthing
工作空间　working space
工作头　working head
工作制　duty
功角　load angle
功率　duty
功率表　wattmeter
功率补偿型差示扫描量热法　power-compensation differential scanning calorimetry
功率计　dynamometer
功率交换　PE (power exchange)
功率谱密度　power spectral density
功率损耗　power loss; watt loss
功率消耗　watt consumption
功率因数表　power factor meter
攻丝夹头　tapping attachment
攻丝装置　tapping attachment
供电　feed
供电点　feeding point
供电电缆　incoming cable
供电柜　incoming cubicle
供电开关　incoming circuit-breaker
供电设备　incoming unit
供电位置　feeding point
供电系统〖经变压器的〗　transformer-fed system
供电线　feeder
供电中断　interruption of power supply
供风道　inlet air duct
供给装置　feeder
供能　feed
供热系统水击　water hammer of heat-supply system
供水　water delivery; water supply
供水管　water supply pipe
汞干电池　mercury cell
汞弧　mercury arc
汞弧变流器　mercury arc converter
汞弧阀　mercury arc valve
拱座　abutment; skewback
共聚物　copolymer
共鸣　resonance
共振　resonance
沟　gutter
构架　carcass
构型　configuration
孤立电厂　isolated power plant
孤立电子　lone electron
股线间短路　short-circuit between strands
骨架　grid
鼓风　blast
鼓风机　blower
鼓轮　drum
鼓形分接头选择开关　tap-selector drum switch
固定电阻器　invariable resistor
固化　solidification
固溶体半导体　solid solution semiconductor
固体　solid
固体放射性废物　solid radioactive waste
固有磁距　intrinsic magnetic moment
固有反馈　inherent feedback
固有负载　inherent burden
固有可靠性　inherent reliability
固有失效　inherent weakness failure
固有衰减　intrinsic damping; natural attenuation
固有瞬态稳定性　inherent transient stability
固有误差　inherent error; intrinsic error
固有延迟　inherent delay

固有周期　natural period
故障　fault
故障安全　fail-safe
故障安全联锁　fail-safe interlock
故障保护设备　fault protective device
故障承受能力　fault withstandability
故障处理　fault handling
故障点　fault point
故障电流　fault current
故障电位　fault potential
故障回路　fault loop
故障检测　fault detection
故障屏蔽　fault masking
故障区段　fault section
故障弱化方式　fail-soft mode
故障探测　fault detection
故障相　fault phase
故障原因　failure cause
故障诊断　failure diagnosis
故障指示器　obstacle indicator
故障阻抗　fault impedance
刮板　scraper
刮刀　scraper
挂起　suspension
挂瓦条　batten
拐点　knee
拐点电压　knee-point voltage
拐点频率　knee frequency
关断间隔　hold-off interval
关合操作　making operation
关合电流　making current
关键帧　key frame
关态电阻　off resistance
观测　observation
观测可靠度　observed reliability
观测频率　observation frequency
观测器　observer
观察　observation
观察孔　inspection hole
观察器　observer
观察箱　inspection box
管簇　tube bundle
管道　piping
管道及仪表流程图　P & ID (piping and instrument diagram)
管道系统　ductwork; pipework
管盖　tube closure
管接头　tube connector
管径　calibre
管连接　tube connection
管路　pipework
管帽　tube cap
管排　tube array; tube bank
管卡　tube clip
管式充油电缆　oil-filled pipe-type cable
管式电缆　pipe-type cable
管式极板　tubular plate
管式加热器　tubular heater
管束　tube bank; tube bundle
管网　piping
管系　piping
管形导体　tubular conductor
管形导线　tubular conductor
管形放电灯　tubular discharge lamp
管形线圈　tubular coil
管涌　piping
管组　tube bank
管座　base
贯流式水轮机　tubular turbine
贯流式通风机　tangential fan
贯入度　penetration
贯入件　penetration
惯量　inertia
惯性　inertia
惯性补偿　inertia compensation
惯性常数　inertia constant
惯性力矩　inertia torque
光泵　optical pump
光编码器　optical encoder
光标　cursor
光标记读取器　OMR (optical mark reader)
光标式仪表　instrument with optical index
光波导　optical waveguide
光参量振荡器　OPO (optical parametric oscillator)
光测高温计　ardometer

光程差 OPD (optical path difference)
光导 light guide
光导传输 light guide transmission
光导管 light conductor
光导衰减 light guide damping
光导体 optical conductor
光电 photoelectricity
光电倍增管 multiplier phototube
光电池 photoelectric cell
光电导性猝熄 quenching of photoconductivity
光电分子 electrogen
光电管 phototube
光电光度计 electrophotometer
光电流 light current
光电子 photoelectron
光电子放大器 optoelectronic amplifier
光电子集成电路 OEIC (optoelectronic IC)
光电子芯片 optoelectronic chip
光电子学 optronics
光电阻 light resistance
光碟头 optical pickup
光度计 photometer
光放大器 light amplifier
光杠杆 optical lever
光隔离器 opto-isolator
光管 light pipe
光集成电路 optical integrated circuit
光继电器 light relay; optical relay
光洁度 finish
光洁度符号 finish mark
光解作用 photolysis
光晶体管 optotransistor
光绝缘体 opto-isolator
光控晶闸管 light-activated thyristor
光雷达 lidar
光雷达跟踪 lidar tracking
光雷达回波 lidar echo
光雷达脉冲 lidar impulse

光量控制器 fader
光量子 light quantum
光敏电阻 LDR (light-dependent resistor); LSR (light-sensitive resistor)
光敏管 LST (light-sensitive tube)
光耦合 optical coupling
光耦合器件 OCD (optically coupled device); opto-coupler
光偏转 light deflection
光偏转器 optical deflector
光频(率) light frequency
光谱 spectrum
光谱半宽度 spectral half width
光谱背景 spectral background
光谱发射率 spectral emissivity
光谱范围 spectral range
光谱分布曲线 spectral distribution curve
光谱辐射计 spectroradiometer
光谱辐射亮度 spectral radiance
光谱化学分析 spectrochemical analysis
光谱能量分布 spectral distribution of energy
光谱特性曲线 spectral characteristic curve
光谱位置 spectral position
光谱狭缝宽度 spectral slit width
光谱线 spectral line
光谱学 spectroscopy
光谱仪 spectrometer
光启硅开关 LASS (light-activated silicon switch)
光启开关 LAS (light-activated switch)
光启可控硅整流器 LASCR (light-activated SCR)
光束分束器 optical beam splitter
光探测 photodetection
光调制器 light modulator; optical modulator
光纤分布式数据接口 FDDI (fibre distributed data interface)

光纤缓冲层 fibre buffer
光纤集中器 optical fibre concentrator
光纤接头 fibre joint; optical fibre splice
光纤耦合器 fibre coupler
光纤散射 optical fibre scattering
光纤束 fibre bundle
光纤束护套 fibre bundle jacket
光纤通信 optical fibre communication
光纤轴 fibre axis
光纤总线 fibre-optic bus
光学 optics
光学安培计 optical ammeter
光学标记读取器 OMR (optical mark reader)
光学成像 optical image formation
光学传感器 optical pickup
光学存储(器) optical storage
光学复用器 optical multiplexer
光学环行器 optical circulator
光学晶体 optical crystal
光学雷达 optical radar
光学模 optical mode
光学谐振 optical resonance
光学字符读取器 optical character reader
光学字符识别 OCR (optical character recognition)
光压 light pressure
光照量 quantity of illumination
光致动器 optical actuator
光中子 photoneutron
光注入 optical injection
光子 photon
光子铅板 Lymar
光字符读取器 optical character reader
光字符识别 OCR (optical character recognition)
归算平板仪 reducing plane table equipment
归一化 normalization
归一化点 normalization point
归一化电阻 normalized resistance
归一化频率 normalized frequency
归一化因子 normalization factor
规度 normality
规格化 normalization
规格化正交性 orthonormality
硅集成 silicon integration
硅酸盐 silicate
硅碳棒 Elema
轨道 orbit
轨道电路 track circuit
轨道电子 orbital electron; planetary electron
轨道函数 orbital function
轨道衡 track scale
轨道回流电路 track return circuit
轨道继电器 track relay
轨道角动量 orbital angular momentum
轨迹 locus
轨距尺 track gauge
滚刀 hob
滚铣 hob
锅壳式锅炉 shell boiler
锅炉给水泵 BFP (boiler feed pump)
锅炉给水系统 BF system (boiler feed system)
锅炉体 boiler vessel
锅筒 drum
国际单位 SI unit
国际单位制 international system of units
过程测量 process measurement
过程定序顺序控制 process-oriented sequential control
过程控制 process control
过程模型 process model
过程输入输出 process I/O
过程责任 process liability
过充电 overcharge
过电流 overcurrent

过电流保护 overcurrent protection
过电流保护配合 overcurrent protective coordination
过电流闭锁装置 overcurrent blocking device
过电流分流器 overcurrent diverter
过电流继电器 overcurrent relay
过电流鉴别 overcurrent discrimination
过电流脱扣器 overcurrent release
过电位 overpotential
过电压 overvoltage
过电压保护 overvoltage protection
过电压抑制器 overvoltage suppressor
过渡触头 transition contact
过渡带 transition band
过渡电流 transition current
过渡电阻器 transition resistor
过渡沸腾 transient boiling; transition boiling
过渡过程曲线 transient curve
过渡环 transition ring
过渡接头 transition joint
过渡区 transient zone; transition region
过渡曲线 transition curve
过渡态 transition state
过渡系数 transition coefficient
过渡系统 intermediate system
过渡线圈 transition coil
过渡阳极 transition anode
过渡状态 transition condition
过渡阻抗 transition impedance
过范围保护 overreaching protection
过负荷 overload
过复励 overcompounding
过复励电动机 overcompounded motor
过激励 over-excitation
过冷却 subcooling
过励磁 over-excitation
过滤 strain
过滤器 filter; strainer
过滤效率 filtration efficiency
过耦合 over-coupling
过耦合变压器 over-coupled transformer
过偏压 overbias
过热 overheat; superheating
过热保护 overheating protection
过热继电器 overheat relay
过热器 superheater
过容量 overcapacity
过压继电器 overvoltage relay
过压力隔离器 overpressure disconnector
过氧化物 peroxide
过载 overload
过载断路器 OCB (overload circuit-breaker)
过载阀 overload valve
过载继电器 overload relay
过载容限 overload margin
过载跳闸 overload trip
过载裕度 overload margin
过阻尼 overdamping

含量 content
焊剂 flux
焊接处 commissure
焊接翻转机 tilter
焊料 solder
焊盘 pad
焊条夹 electrode arm
焊趾裂纹 toe crack
焊珠 bead
夯 tamper
夯实 compaction
夯实回填土 tamped backfill
夯实炉渣 tamped cinder
行频 horizontal frequency
行扫描 line scanning
航路 airway
航向信标 localizer
毫安 ma (milliampere)
毫安表 milliammeter
毫伏计 millivoltmeter
毫伏(特) mv (millivolt)
毫克 mg (milligram)
毫米 mm (millimetre)
毫姆(欧) millimho
毫欧表 milliohmmeter
毫欧(姆) milliohm
毫升 ml (millilitre)
毫瓦(特) mw (milliwatt)
毫微安计 nanoammeter
毫微安(培) na (nanoampere)
毫微程序 nanoprogram
毫微处理器 nanoprocessor
毫微法 nanofarad
毫微伏(特) nanovolt
毫微亨(利) nanohenry
毫微加工技术 nanotechnology
毫微米 millimicron; nm (nanometre)
毫微秒 ns (nanosecond)
毫微瓦(特) nanowatt
耗功 wasted work
耗尽层 depletion layer
耗散 dissipation
合成代谢 anabolism
合格品 non-defective
合金 alloy
合路器 combiner; mixer
合闸瞬态电流 inrush transient current
核磁共振 NMR (nuclear magnetic resonance)
核电 nuclear power
核电荷 nuclear charge
核激光器 nuclear laser
核石墨 nuclear graphite
核子 nucleon
荷 charge
赫维赛德效应 Heaviside effect
黑沥青 abbertite
痕量分析 trace analysis
恒磁通调压 constant flux voltage regulation
恒温器 thermostat
恒值调节器 isodromic governor
桁架式塔架 lattice tower
横差保护 transverse differential protection
横场 transverse field
横磁波 transverse magnetic wave
横磁化 transverse magnetization
横电磁波 transverse electromagnetic wave
横断面 cross-section; transverse section

横截面 cross-section; transverse section
横联差动保护 transverse differential protection
横流扇 tangential fan
横向布置 transverse arrangement
横向布置凝汽器 transverse condenser
横向槽 transverse slot
横向场 transverse field
横向磁场 transverse magnetic field
横向磁化 transverse magnetization
横向磁通加热 transverse flux heating
横向断裂 transverse fracture
横向分辨力 transverse resolution
横向钢筋 transverse reinforcement
横向构架 bent
横向混合 transverse mixing
横向节距 transverse pitch
横向晶体管 lateral transistor
横向聚焦 transverse focusing
横向绝缘 lateral isolation
横向力 transverse force
横向流 transverse flow
横向偏转 lateral deflection
横向强度 transverse strength
横向曲率 cross curvature
横向屈曲 transverse buckling
横向伸缩缝 transverse expansion joint
横向压电效应 transverse piezoelectric effect
横向衍射 transverse diffraction
横向应变 transverse strain
横向应力 transverse stress
横向运动 transverse movement
横向振荡 transverse oscillation
横向振动 transverse vibration
横楔 transverse wedge
横轴 transverse axis
横坐标 abscissa; x-coordinate
轰击 bombardment
红外单色仪 infrared monochromator
红外辐射 infrared radiation
红外光谱法 infrared spectroscopy
红外光谱学 infrared spectroscopy
红外探测器 infrared detector
红外温度计 infrared thermometer
红外吸收 infrared absorption
红外线 ultrared ray
红外遥控器 infrared controller
红外遥控系统 infrared remote control system
红外中央控制闭锁系统 infrared central locking system
宏单元 macrocell
宏电路 macrocircuit
宏观常数 macroscopic constant
宏观应力 macroscopic stress
宏块 macroblock
宏指令 macro-instruction
虹吸管 siphon; syphon
后冷凝器 after-condenser
后冷器 aftercooler
后倾式电刷 trailing brush
后效时间 impression time
后沿延迟 trailing-edge delay
后缘 TE (trailing edge)
厚板 plank
厚薄规 feeler
弧 arc
弧度 radian
弧光灯驱动器 arc lamp drive
弧焊 arc welding
互变 interconversion
互补性 complementarity
互导纳 transadmittance
互感器 instrument transformer
互换 interconversion
互换电路 interchange circuit
互换功率 interchange power
互接掩模 interconnection mask

互抗 transreactance
互连 interconnection
互连运行 interconnected operation
互连运转 interconnected operation
互纳 transsusceptance
互调 intermodulation
互阻 transresistance
互阻抗 transimpedance
互作用场 interaction field
互作用隙 interaction gap
户内变电站 indoor substation
户内套管 indoor bushing
户内外绝缘 indoor external insulation
护盖 bonnet
护墙板 sheeting
护套 sheath(ing)
护舷材 fender
护罩 bonnet
花键 spline
滑板 slider
滑车 tackle
滑车轮 sheave
滑道 shoot
滑底式炉 skid hearth furnace
滑动 sliding
滑动触头 sliding contact
滑动接触 sliding contact
滑动块 slider
滑阀 slide valve
滑过触点 passing contact
滑环 slip ring
滑块 slider
滑轮 tackle block
滑线 skid wire
滑移 sliding
滑闸 shuttle
化能合成 chemosynthesis
化学发光 chemiluminescence
化学计量 stoichiometry
划针 scriber
还原黏度 reduced viscosity
环流 circulation
环流电动机 loop motor

环流量 circulation
环路 loop
环路增益 LG (loop gain)
环式馈线供电单元 incoming ring-feeder unit
环烷烃 cyclane
环芯变压器 toroidal transformer
环行器 circulator
环形存储磁芯 toroidal memory core
环形空腔谐振器 toroidal cavity resonator
环形绕组 toroidal winding
缓波前过电压 slow-front overvoltage
缓冲存储器 elastic store
缓冲垫 cushion
缓冲器 absorber; buffer; snubber
缓动 inching
缓动继电器 time-delay relay
缓动速度 inching speed
缓蚀剂 inhibitor
换接母线 transfer bus
换流变压器 converter transformer
换流器 converter; transverter
换流器电路识别码 identification code for converter connections
换能器 transducer
换气率 ventilation rate
换热器 heat exchanger
换热系数 heat exchange coefficient
换位 transposition
换位导线 transposed conductor
换位间隔 transposition interval
换位矩阵 transposed matrix
换位绝缘子 transposition insulator
换位绕组 transposed winding
换位塔 transposition tower
换位循环 transposition cycle
换向阀 reversing valve
换向极 interpole
换向极绕组 interpole winding

换向器 commutator
黄铁矿 pyrite
黄铜 brass
灰坑 ash pit
回波 echo
回波衰减 echo attenuation
回波箱 echo box
回波像 echo-image
回波消除器 echo eliminator
回风风口 return outlet
回风管道 return air duct
回复值 reverting value
回话电路 talk-back circuit
回流电缆 return cable
回流电路 return circuit
回流换热器 recuperator
回路 circuit; loop
回路电流 loop current
回热循环 regenerative cycle
回声 echo
回声测深图 echogram
回声测深仪 echo sounder
回声定位 echolocation
回声深度记录 echogram
回声仪 echometer
回声抑制器 echo trap
回授线圈 tickler (coil)
回水阀 water return valve
回水管 water return pipe
回送 echo
回跳硬度计 Scleroscope
回旋质谱计 omegatron
回转压缩机 rotary compressor
汇 sink
汇编程序 assembler
汇点 sink
汇接中继机键 tandem switch
汇流阀 manifold valve
汇流排 busbar
惠斯通电桥 Wheatstone bridge
混床 mixed bed
混合 hybrid
混合床 mixed bed
混合电路 hybrid circuit
混合多用表 hybrid multimeter
混合模块 hybrid module
混合配置 hybrid configuration
混合器 commingler
混合驱动 hybrid drive
混合式不间断电源开关 hybrid UPS power switch
混合式连接器 hybrid connector
混合式适配器 hybrid adaptor
混合式转接器 hybrid adaptor
混合数据采集系统 hybrid data acquisition system
混合调压 combined voltage regulation
混合万用表 hybrid multimeter
混合系统 hybrid system
混合子 hybrid
混频二极管 mixer diode
混频器 mixer
混水连接 water-mixing direct connection
混水装置 water admixing installation
活动测高计 travelling cathetometer
活度 activity
活化剂 activator
活节螺栓 eye bolt
活塞 piston
活性 activity
活性剂 activator
火管锅炉 shell boiler
火花点火 spark ignition
火花检测器 spark tester
火花塞 spark plug
火花塞抑制器 spark plug suppressor
火箭探空仪 rocketsonde
火炬点火器 torch igniter
火山灰水泥 trass cement
火焰弧光灯 flame arc lamp
火焰式电晕 torch corona
或 OR
或电路 OR circuit
或非电路 NOR circuit
或非门 NOR gate
或功能 OR function
或函数 OR function

或开关　OR switch
或门　OR gate
或门管　OR tube
或算子　OR operator
或元件　OR element
获能腔　catcher resonator
霍尔板　Hall plate
霍尔电压　Hall voltage
霍尔发生器　Hall generator
霍尔迁移率　Hall mobility
霍尔探头　Hall probe
霍尔调制器　Hall modulator
霍尔效应　Hall effect
霍尔效应传感器　Hall-effect sensor
霍尔效应磁强计　Hall-effect magnetometer
霍尔效应开关　Hall-effect switch
霍尔效应器件　Hall-effect device
霍尔效应探头　Hall-effect pickup

击穿电压 puncture voltage
机床 machine tool
机电换能器 electromechanical transducer
机电继电器 electromechanical relay
机电模拟 electromechanical analogy
机电耦合 electromechanical coupling
机电器件 electromechanical device
机电学 electromechanics
机构 mechanism
机壳 housing
机理 mechanism
机器 apparatus; machine
机箱 cabinet
机械测振法 vibration measurement by mechanical method
机械电子技术 mechatronics
机械电子学 mechatronics
机械加工性 machinability
机械破坏负荷 mechanical failure load
机械学 mechanics
机械转矩率 mechanical torque rate
机制 mechanism
机组降额发电量 UDG (unit derating generation)
机组降额因数 UDF (unit derating factor)
机组转入单机运行 isolation of a unit
奇对称 odd symmetry
奇函数 odd function
奇调和函数 odd harmonic function
积层磁铁 laminated magnet
积分控制 integral control
基本绝缘水平 BIL (basic insulation level)
基础灌浆 foundation grouting
基带 normal band
基尔霍夫电压定律 KVL (Kirchhoff's voltage law)
基尔霍夫定律 Kirchhoff's law
基极 base
基坑 foundation pit
基体 matrix
基线 baseline
基准 benchmark
基准孔 indexing hole
基准线圈 reference coil
基准值 fiducial value
基座坐标系 base coordinate system
激磁装置 magnetizer
激发 excitation
激光 laser
激光变像器 LIC (laser image converter)
激光干扰滤波器 LIF (laser interference filter)
激光光学调制器 LOM (laser optical modulator)
激光焊接 laser welding
激光雷达 lidar
激光雷达跟踪 lidar tracking
激光雷达回波 lidar echo
激光雷达脉冲 lidar impulse
激光器 laser

激光闪光管 LFT (laser flash tube)
激光转换器 lasecon
激活剂 activator
激励 excitation
激励猝灭 quenching of excitation
激励电压 energizing voltage
激励管 excitron
激励频率 energizing frequency
激励器 driver; exciter
激励设备 energizing apparatus
激励位置 energized position
激励状态 energized condition
激振 excitation
激振器 (vibration) exciter; vibration generator; vibrator
级 stage
级间变压器 interstage transformer
级间耦合 interstage coupling
级联电动机 tandem motor
极低 LL (low-low)
极低频 ELF (extremely low frequency)
极高频 EHF (extremely high frequency)
极高压 EHT (extra-high tension); EHV (extra-high voltage)
极化电荷 polarization charge
极间电容 electrode capacitance
极距 pole pitch
极面 pole face
极限 pH ultimate pH
极限变形 ultimate deformation
极限承载力 ultimate bearing capacity
极限分辨率 limiting resolution
极限分断容量 limiting breaking capacity
极限负荷 ultimate load
极限工作容量 ultimate working capacity
极限机械强度 ultimate mechanical strength
极限抗拉强度 ultimate tensile strength
极限抗拉应力 ultimate tensile stress
极限抗压强度 ultimate compressive strength
极限拉伸强度 ultimate tensile strength
极限拉应力 ultimate tensile strength
极限强度 ultimate strength
极限强度设计 ultimate strength design
极限曲线 limit curve
极限容量 limit capacity; ultimate capacity
极限伸长 ultimate elongation
极限衰减 ultimate damping
极限弯矩 ultimate bending moment
极限误差 limiting error
极限应变 ultimate strain
极限载荷 limit load; ultimate load
极限载荷设计 ultimate load design
极限张力 ultimate tension
极限真空 ultimate vacuum
极限值 limit value; ultimate value
极限阻力 ultimate resistance
极限阻尼 ultimate damping
极限最大值 ultimate maximum
极性 polarity
极靴 pole shoe
集成电路 IC (integrated circuit); MC (microcircuit)
集成电路继电保护系统 integrated-circuit relay protection system
集成光路 integrated optical circuit
集成计算机辅助制造 integrated computer-aided manufacturing
集成空气/燃油系统 integrated air/fuel system

集成微电路 integrated microcircuit
集电环 slip ring
集电极 collector
集肤效应 Heaviside effect
集汽管 steam receiver
集水坑 sump
集水区 water catchment area
集线器 concentrator; hub; LC (line concentrator)
集箱 header
集油槽 sump
集中参数电路 lumped circuit
集中器 concentrator
几何畸变 geometric distortion
几何失真 geometric distortion
挤包绝缘 extruded insulation
挤水棒 WDR (water displacer rod)
给水泵 water supply pump
给水工程 water supply engineering
给水器 water feeder
给水系统 water supply system
计 meter (m)
计尘试验 konitest
计量送料器 batcher
计量型检查 inspection by variables
计量学 metrology
计时电容器 timing capacitor
计时器 timer
计数器 numerator
计数型触发器 toggle flip-flop
计数型检查 inspection by attributes
计算机辅助设计 CAD (computer-aided design)
计算机辅助生产 CAP (computer-aided production)
记波法 kymography
记波器 kymograph
记波图 kymogram
记录 logging
记录器 logger
技术检验报告 technical inspection report
技术设计 technical design
技术数据 TD (technical data)
技术系数 technical coefficient
剂量 dose
剂量当量 dose equivalent
剂量计 dosimeter
继承误差 inherited error
继电保护 relaying protection
继电器 relay
加标 labelling
加感电缆 LC (loaded cable)
加感线圈 loading coil
加工岛 island of production
加工符号 finish mark
加工过程检测 in-process inspection
加劲杆 stiffener
加劲肋 stiffener
加力燃烧 afterburning
加力燃烧室 afterburner
加氯处理 chlorination
加煤机 stoker
加气 AE (air entraining)
加气混凝土 aerated concrete
加强条 stiffener
加热导体 heating conductor
加热电缆单元 heating cable unit
加热电路 heating circuit
加热电阻体 heating resistor
加伸轴 stub shaft
加载 loading
加载共振器 loaded resonator
夹层 interlayer
夹带水分 water carry-over
夹件 clamping device
夹具 jig
夹线式电流表 tong-test ammeter
夹子 keeper
家用电器 household (electrical) appliance
家用电子系统应用协议 HES application protocol (home electronic system application protocol)
甲级板材 first-grade sheet

甲烷馏除器 demethanizer
钾 potassium
假象效应 Kendall effect
假信号 glitch
架承式电动机 frame-mounted motor
架空导线 overhead conductor
架空电缆 overhead cable
架空线路 overhead line
尖峰负荷电站 peak-load plant
尖峰态 leptokurtosis
尖峰信号 spike
尖头插孔 tip jack
间距 gap; spacing
监视电路 observation circuit
兼容性 compatibility
检波器 pickup; wave detector
检查 inspection
检查井 manhole
检查批 inspection lot
检查箱 inspection box
检流计 galvanometer
检漏仪 leakage detector; leak detector
检修接地 inspection earthing
检验 inspection
检验报告 inspection report
检验标签 inspection sticker
检验量规 inspection gauge
减负荷 deload
减水外加剂 water-reducing admixture
减温器 attemperator; desuperheater
减压阀 pressure reducing valve
减压器 decompressor
减震器 absorber; buffer; damper; snubber
减阻装置 fairing
剪切 shear
剪切机 shear
碱 alkali; base
碱度 alkalinity; basicity
碱化 alkalinization
间断接地故障 intermittent earth fault
间断性电弧接地 intermittent arcing ground
间断运行 intermittent operation
间隔 interval; spacing
间隔棒 spacer
间隔时间计数器 interval counter
间接点火 indirect firing
间接电加热 indirect electric heating
间接过流脱扣器 indirect overcurrent release
间接空气冷却电机 indirectly air-cooled machine
间接雷击 indirect lightning strike
间接冷却绕组 indirect cooled winding
间接驱动电机 indirect-drive machine
间接水冷式空冷电机 water-air-cooled machine
间接作用式仪表 indirect-acting instrument
间隙 gap; interval
间隙规 gap gauge
间隙绕包 open lapping
间歇电流 intermittent current
间歇故障 intermittent fault
间歇失效 intermittent failure
间歇运转 intermittent duty
间歇运转测量 intermittent rating
间歇运转测量值 intermittent rating
间歇运转功率 intermittent rating
建立时间 settling time
渐变截面叶片 tapered blade
渐减曝气 tapered aeration
渐近速度系数 velocity of approach factor
渐开线连接 involute connection
渐开线型绕组 involute winding
鉴定标准 qualification standard
鉴定试验 qualification test

鉴时器 time discriminator
键 key
键棒 keybar
键槽 key bed; key seat; keyway
键架 key shelf
键控 key
键控穿孔 keypunch
键控穿孔机 keypunch
键控电路 keying circuit
键控法发报 key
键控管 key tube
键控继电器 key relay
键控脉冲 KP (key pulse)
键控频率 keying frequency
键控器 keyer
键控信号 keying signal
键控信号波 keying wave
键盘 keyboard; keyset
键盘编码器 key coder
键盘处理器 keyboard processor
键盘开关 keyboard switch
键区 keypad
键式开关 key switch
浆体 slurry
降级 degradation
降阶观测器 reduced order observer
降阶模型 reduced model
降解 degradation
降容量分接 reduced power tapping
降压变压器 step-down transformer
降压阀 relief valve
交 intersection
交叉棒 cross bar
交叉段 transposition section
交叉间隔 transposition interval
交叉绝缘子 transposition insulator
交叉连接 cross-connection
交叉区 transposition section
交叉塔 transposition tower
交叉支撑 X-brace
交错结合点 interleaved joint
交错阻抗电压 interlacing impedance voltage
交点 intersection
交互方式 interactive mode
交互工作单元 interworking unit
交互式处理 interactive processing
交互作用 interaction
交换功率 interchange power
交换功能 interchange function
交换机 exchanger
交换剂 exchanger
交换器 exchanger
交换子 commutator
交接电流 takeover current
交链磁通 interlinked flux
交链漏磁通 interlinked leakage flux
交链系数 interlinking factor
交流 AC (alternating current)
交流电流 AC (alternating current)
交流电子开关 electronic AC switch
交流发电机 alternator
交流声 hum
交直流两用电动机 universal motor
交轴分量 quadrature-axis component
交轴绕组 quadrature-axis winding
胶浸纸套管 resin-impregnated paper bushing
胶黏剂 adhesive
焦耳效应 Joule effect
焦化值 tarring number; tarring value
焦油砂 tar sand
角 angle
角撑 knee bracing
角度 angle
角焊缝 fillet weld
角式燃烧器 tangential burner
角行程气动执行机构 pneumatic angular displacement actuator

绞合导体 stranded conductor
绞合导线 stranded conductor
绞合接头 twist joint
绞合软线 twisted cord
绞接 twist joint
绞接头 twisted joint
绞距 length of lay
绞线 stranded conductor; twisted line
绞线换位 twisted-lead transposition
铰接 articulation
矫直机 straightener
脚球 pin ball
搅拌 agitation; stirring
搅拌器 agitator; stirrer
搅动 agitation
校平器 leveller
校正器 corrector
校正子 corrector
校准 calibration
校准器 calibrator
阶梯式迷宫式汽封 stepped labyrinth gland
接地 earth (connection); earthing; ground
接地板 EP (earth plate); grounding pad
接地板电阻 earth plate resistance
接地棒 earth bar; earth rod
接地变压器 earthing transformer
接地测量仪 earthometer
接地差动保护 ground differential protection
接地插孔 earth jack
接地插头 ground connector
接地出线 grounding outlet
接地触点 earthing contact
接地带 earth strip
接地刀 grounding blade
接地导体 earthing conductor; ground conductor
接地导线 earth conductor; earth lead
接地点 earth point

接地电极 earth electrode; grounding electrode
接地电抗器 earthing reactor
接地电缆 grounding cable
接地电流均衡器 ground-current equalizer
接地电流限制器 ground-current limiter
接地电路 earth circuit
接地电刷 earthing brush
接地电压 earthed voltage
接地电阻 earth resistance; earth resistor; ground resistance
接地垫 grounding pad
接地短路测定 ground-fault location
接地返回电路 ground return
接地故障 earth fault; ground fault
接地故障保护 earth-fault protection; ground-fault protection
接地故障补偿 ground-fault compensation
接地故障电弧 ground-fault arc
接地故障电抗器 ground-fault reactor
接地故障电流 earth-fault current; ground-fault current
接地故障电路保护 ground-fault circuit protection
接地故障电路中断器 ground-fault circuit interrupter
接地故障回路阻抗 ground-fault loop impedance
接地故障继电器 earth-fault relay; ground-fault relay
接地故障监视器 ground-fault monitor
接地故障试验 ground-fault test
接地故障探测器 ground-fault detector
接地故障位置 ground-fault location
接地故障中和器 ground-fault neutralizer

接地回流电路 earth return circuit
接地回路 earth loop; grounded circuit
接地回线 ER (earth return)
接地汇流排 ground bus
接地极 earth electrode; grounding electrode
接地继电器 earth-fault relay; earthing relay
接地夹 earth clip; ground clamp
接地检测器 ground detector
接地接头 ground connector; ground joint
接地绝缘 ground insulation
接地开关 earth(ing) switch; grounding switch
接地漏电 earth leakage
接地漏电断路器 ELCB (earth leakage circuit-breaker)
接地母线 earth bus; ground bus
接地屏蔽 (earth) shield
接地栅极 earth grid
接地探测器 ground detector
接地体 earth electrode
接地跳线 grounding jumper
接地系统 earthing system
接地线 earth connection; ground wire
接地线棒 grounding bar
接地线夹 earth clamp
接地型插头 grounding-type plug
接地型插座 grounding-type receptacle
接地因数 factor of earthing
接地指示器 ground indicator
接地中性点 grounded neutral
接地装置 ground
接缝 commissure; joint
接合 joint
接合处 commissure
接口 interface
接口分配 interface assignments
接口管理器 interface handler
接口管理总线 interface management bus
接口协议 interface protocol
接口主体 interface agent
接口总线 interface bus
接通 interface
接通操作 making operation
接通电流 making current
接通电流峰值 inrush peak
接通开关 making switch
接通能力 making capacity
接通涌流 inrush making current
接通状态 on-state
接头 collar; connector; joint; junction
接头焊接 joint welding
接线盒 joint box; terminal box
接线片 lug
接线图 hook-up
接线箱 junction box
节 knot
节点导纳矩阵 nodal admittance matrix
节点电压 node voltage
节点法 node method
节点力 nodal force
节间 panel
节距因数 pitch factor
节流门 strangler
节流器 restrictor
节流损失 throttling loss
节圆 nodal circle
节圆直径 pitch diameter
结 junction
结构分解 structural decomposition
结构高度 spacing
结构公式 structural formula
结构化程序 structured program
结构化程序设计 structured programming
结构化程序设计语言 structured programming language
结构可观测性 structural observability
结构可控性 structural controllability
结构可通性 structural passability
结构模型 structure model

结构能观测性 structural observability
结构能控性 structural controllability
结构能通性 structural passability
结构式 structural formula
结构协调 structural coordination
结果程序计算机 target computer
结合器 adaptor
结晶 crystal
结晶水 water of crystallization
结晶体 crystal
结块 caking
结露 dewing
结算指令 tally order
结线 tie line
结型晶体管 junction transistor
截断瞬间 instant of chopping
截止阀 stop valve
截止晶体管 off transistor
截止状态 off condition
解除联锁钥匙 interlock deactivating key
解除联锁装置 interlock deactivating means
解环 looping-off
解列 splitting
解码器 decoder
解絮凝 deflocculation
解絮凝剂 defloccoulant
介电常数 dielectric constant; inductivity
介电常数倒数 elastivity
介电体 dielectric
介子 meson
界面 interface
界面反应率常数 interface reaction-rate constant
界面管理器 interface handler
界面效应 interface effect
界限系数 marginal coefficient
金具 fitting
金属箔电容器 metal foil capacitor
金属弧焊 metal arc welding
金属护套 metallic sheath
金属化 metallization
金属回线 metallic return
金属键 metallic bond
金属绝缘体 metal insulator
金属膜电阻器 metal film resistor
金属氧化物半导体 MOS（metal oxide semiconductor）
金斯伯里轴承 Kingsbury bearing
紧带轮 idler
紧定螺钉 set screw
紧急停堆 scram
紧耦合 tight coupling
紧套光纤 tight tube fibre
进刀 feed
进刀机构 feeder
进刀驱动 feed drive
进风 air intake
进给 feed motion; infeed
进给量 feed rate
进给驱动 feed drive
进给运动 feed motion; infeed motion
进给轴 infeed axis
进口 inlet
进料 feed
进气道 air intake
进气压力 inlet pressure
进汽箱 steam chest
进入馈线 incoming feeder
进入馈线间隔 incoming-feeder bay
进水口 water inlet; water intake
进水效应 water logging effect
进线 incoming feeder
进线间隔 incoming-feeder bay
进线避雷线 incoming overhead ground wire
进线配电盘 incoming panel
近端串音 near-end crosstalk
近红外 NIR（near infrared）
近回波 near echo
近区故障 SLF（short-line fault）

近紫外 NUV (near ultraviolet)
劲度 stiffness
浸漆织物 varnished fabric
禁带 forbidden band
晶格常数 lattice constant
晶格畸变 lattice distortion
晶格空位 lattice vacancy
晶格匹配 lattice matching
晶格缺陷 lattice defect
晶格振动 lattice vibration
晶核 nucleus
晶石 spar
晶体 crystal
晶体磁控管 madistor
晶体管 transistor
晶体管参数 transistor parameter
晶体管测量示波器 transistor measuring oscilloscope
晶体管触发器 transistor flip-flop; transistor trigger
晶体管存储器 transistor memory
晶体管电动机 transistor motor
晶体管电流导引逻辑 TCSL (transistor current steering logic)
晶体管电路 transistor circuit
晶体管电压表 transistor voltmeter
晶体管电压稳定器 transistor voltage stabilizer
晶体管电子学 transistor electronics
晶体管-电阻器-晶体管逻辑 transistor-resistor-transistor logic
晶体管-电阻器逻辑 transistor-resistor logic
晶体管多谐振荡器 transistor multivibrator
晶体管-二极管-晶体管逻辑 transistor-diode-transistor logic
晶体管-二极管逻辑 transistor-diode logic
晶体管方程 transistor equation
晶体管放大器 transistor amplifier
晶体管分析记录设备 TARE (transistor analysis recording equipment)
晶体管伏特计 transistor voltmeter
晶体管及元件测试仪 TACT (transistor and component tester)
晶体管集电极 transistor collector
晶体管继电器 transistor relay
晶体管解调器 transistor demodulator
晶体管开关 transistor switch
晶体管逻辑电路 transilog; transistor logic circuit
晶体管耦合晶体管逻辑 TCTL (transistor-coupled transistor logic)
晶体管耦合逻辑 TCL (transistor-coupled logic)
晶体管偏压电路 transistor bias circuit
晶体管十进制计数器 transistor decade counter
晶体管式继电器 transistor relay
晶体管伺服前置放大器 transistor servo preamplifier
晶体管稳压器 transistor voltage stabilizer
晶体管显示器及数据处理系统 transistor display and data processing system
晶体管斩波器 transistor chopper
晶体管振荡器 transistor oscillator
晶体管转矩仪 transistor torque meter
晶闸管模块 thyristor module
精刨 finite planing
精定位 fine positioning
精加工 finish
精密调整 fine adjustment
精细进给 fine feed
精整度 finish

井道 hoistway
井架 derrick
井式溢洪道 shaft spillway
警报 warning alarm
警告信号 warning signal
警戒界限 warning limit
净电荷 net charge
净发电量 net generation
净宽度 inside width
净气 sweet gas
净输出 net output
净水槽 water purifying tank
净水厂 water purification plant
净水构筑物 water purification structure
净损耗 net loss
静不平衡 static unbalance
静点 quiescent point
静电 static electricity
静电变压器 static transformer
静电场 electrostatic field
静电除尘器 ESP (electrostatic precipitator)
静电发电机 influence machine
静电感应 electrostatic induction
静电计 electrometer
静电加速器 electrostatic accelerator
静电喷粉 electrodusting
静电喷粉器 electroduster
静电偏转 electrostatic deflection
静电屏蔽 electrostatic screening
静电吸引 electrostatic attraction
静电吸引定律 law of electrostatic attraction
静电学 electrostatics
静电印刷机 xerographic printer
静电印刷术 xerography
静动态应变仪 static-dynamic strain indicator
静力弹簧重力仪 static spring gravimeter
静力学 statics
静摩擦 static friction
静扭矩 static torque
静平衡 static balance

静态 statics
静态标准应变装置 static standard strain device
静态测量 static measurement
静态测量法 static gauging method
静态称重法 static weighing method
静态电流 quiescent current
静态二次离子质谱法 static SIMS
静态负荷 static load
静态负荷补偿装置 static load compensating device
静态挂码 static weight-hoist
静态继电器 static relay
静态校准 static calibration
静态解耦 static decoupling
静态精密度 static precision
静态精确度 static accuracy
静态模型 static model
静态热测量仪表 static thermal measuring instrument
静态热技术 static thermal technique
静态特性 static characteristics
静态特性曲线 static characteristic curve
静态显示图像 static display image
静态压力传感器 static pressure sensor; static pressure transducer
静态应变仪 static strain indicator
静态整步转矩 static synchronizing torque
静态质谱仪 static mass spectrometer
静特性曲线 transfer curve
静压 static pressure
静液压泵 hydrostatic pump
静应变 static strain
静转矩 static torque
静子 stay
纠结式双线圈 interleaved double coil

纠结式相绕组 interleaved phase windings
纠结线圈 interleaved coil
酒石酸 tartaric acid
酒石酸溶液 tartaric acid solution
局部电路 local circuit
局部反馈 LFB (local feedback)
局部分布 local distribution
局部停电 brown-out
矩 moment
矩阵 matrix
矩阵电路 matrix circuit
矩阵定理 matrix theorem
矩阵反演 matrix inversion
矩阵加法器 matrix adder
矩阵求逆 matrix inversion
矩阵显示 matrix display
矩阵寻址 matrix addressing
距离 interval
锯齿波 toothed wave
聚沉 coagulation
聚光器 condenser
聚合物 polymer
聚集 agglomerate
聚焦电极 focusing electrode
聚焦线圈 focusing coil
聚酰胺 PA (polyamide)
聚酯树脂 Mylar
卷带机构 tape mechanism
卷积 convolution
卷绕铁芯 wound core
卷轴 reel
绝对测振法 vibration measurement by absolute method
绝对零度 absolute zero
绝对湿度 absolute humidity
绝对温标 Kelvin scale
绝对温度 absolute temperature; Kelvin degree
绝对误差 absolute error
绝对压力 absolute pressure
绝对压强 absolute pressure
绝缘 insulation; isolation
绝缘布 empire cloth
绝缘材料 insulant; insulation
绝缘成本 insulation cost
绝缘带 insulating tape
绝缘等级 insulation class
绝缘电缆 insulated cable
绝缘电阻 insulance
绝缘段 isolating distance
绝缘击穿 insulation breakdown
绝缘夹钳 insulated clamp
绝缘间隙 insulating clearance
绝缘胶 insulating cement
绝缘胶带 insulating tape
绝缘介质 insulating agent
绝缘连接 insulated coupling
绝缘联轴节 insulated coupling
绝缘能力 insulating ability
绝缘丝 empire silk
绝缘体 insulant; insulator; isolator; non-conductor
绝缘涂层 insulating coating
绝缘性质 insulating property
绝缘子 insulator
绝缘子串 insulator string
绝缘子式电流互感器 insulator-type current transformer
均方根 root mean square
均衡 equalization
均衡充电 equalizing charge
均衡器 equalizer
均压环 equalizing ring
均压线 equalizer
菌胶团 zoogloea

喀斯特 karst
卡 calorie
卡尺 caliper
卡尔曼滤波器 Kalman filter
卡尔曼周期 Kalman cycle
卡口灯头 bayonet base; bayonet cap
卡利罗管 kallirotron
卡玛箔 karma foil
卡门涡流 Karman's vortex
卡诺图 Karnaugh map
卡普-霍普金森试验 Kapp-Hopkinson test
卡普兰式水轮机 Kaplan turbine
卡钳 caliper
卡氏系数 Kapp coefficient
开 K (kelvin)
开尔文 K (kelvin)
开尔文秤 Kelvin balance
开尔文电桥 Kelvin bridge
开尔文双电桥 Kelvin double bridge
开尔文温度 Kelvin degree
开尔文效应 Kelvin effect
开关 switch
开关臂 switch arm
开关场 switchyard
开关电子管 nomotron
开关设备 switchgear
开关式电源 SMPS (switch-mode power supply)
开关式供电 SMPS (switch-mode power supply)
开合 switch
开环控制 open-loop control
开环系统 open-loop system
开路 OC (open circuit)
开路电弧 open-circuit arc
开路电压 open-circuit voltage
开路试验 open-circuit test
开路自转 open-circuit spinning
开启式谐振腔 open-type resonator
开氏温标 Kelvin scale
开通 firing
开通阀 firing valve
凯勒氏电弧炉 Keller furnace
凯泽效应 Kaiser effect
抗爆剂 antiknock; knock-inhibiting additive
抗干扰能力 immunity to inference
抗扭配筋 torsional reinforcement
抗扰度 immunity to inference
抗扰度电平 immunity level
抗熔性 non-fusibility
柯达标准片孔 Kodak Standard perforation
柯尔劳希电桥 Kohlrausch bridge
可变标准电容器 variable standard capacitor
可变电感器 variable inductor; variometer
可变电容 variable capacitance
可变电容器 variable capacitor
可变放大系数 variable gain
可变负载 LL (live load)
可变角内反射元件 variable-angle internal reflection element
可变角探头 variable-angle probe
可变耦合器 variocoupler

可变增益 variable gain
可变增益法 variable gain method
可拆元件 removable element
可持续性 sustainability
可充电电池 rechargeable battery
可重接插头 rewirable plug
可抽件 withdrawable part
可磁化性 magnetizability
可服务性 serviceability
可更换熔断体 renewable fuse-link
可观测性 observability
可忽略误差 insignificant non-conformance
可见度 visibility
可见光传感器 visible light sensor; visible light transducer
可见光辐射 visible radiation
可见光遥感 visible spectral remote sensing
可靠度 reliability
可靠性 reliability; soundness
可控硅整流器 SCR (silicon controlled rectifier)
可切削性 machinability
可倾瓦块轴承 tilting-pad bearing
可燃性 combustibility
可溶物质 solvend
可溶性 solubility
可实现性 feasibility
可调变压器 transtat
可调绕组 teaser winding
可维护性 maintainability
可维修度 maintainability
可维修性 maintainability
可行性 feasibility
可再生能源 renewable energy
可中断负荷 interruptible load
可中断负载 interruptible load
克尔盒 Kerr cell
克尔效应 Kerr effect
克努森滴定管 Knudsen's burette
克努森移液管 Knudsen's pipette
刻度 calibration
刻图仪 scriber
客观光度计 objective photometer
肯德尔效应 Kendall effect
空操作指令 no-op instruction
空带 empty band
空放阀 relief valve
空化 cavitation
空距 ullage
空能级 empty level
空泡 cavity
空气间隙 air gap
空气冷却 air-cooling
空气冷却器 air-cooler
空气摩擦损耗 windage loss
空气隙 air gap
空气预热器 AH (air preheater)
空速管 Pitot tube
空心绝缘子 hollow insulator
空心轴电动机 hollow-shaft motor
空心轴型测速发电机 hollow-shaft-type tachogenerator
空行程 idle stroke
空穴存储效应 hole storage effect
空穴导电性 hole-type conductivity
空穴电导率 hole-type conductivity
空穴色谱法 vacancy chromatography
空穴型半导体 hole semiconductor
空载表观功率 no-load apparent power
空载电流 no-load current
空载电路 idle circuit
空载分接开关 no-load tap-changer
空载时间 idle time
空载试验 no-load test
空载损耗 no-load loss
空载损失 no-load loss
空载运行 empty running; no-load operation

空载转速 no-load speed
空转 idling
空转轮 idler
空转时间 idle time
空转速度执行器 idle speed actuator
孔板流量计 orifice flowmeter
孔径 hole diameter
孔口 opening; orifice
孔口出流 orifice flow
孔隙度 porosity
空隙 cavity
空闲容量 idle capacity
空闲状态 idle state
控制变压器 control transformer
控制触头 control contact
控制电路 control circuit
控制电器 control apparatus
控制极 control electrode
控制接线图 CWD (control wiring diagram)
控制精(确)度 control accuracy
控制开关 CS (control switch)
控制励磁机 control exciter
控制论 cybernetics
控制面板 fascia
控制盘 control panel
控制台 console
控制箱 control box
控制准确度 control accuracy
口径 bore; calibre
库 library
库容 content; storage capacity
裤衩管 Y-bend
跨导 transconductance
跨导纳 transadmittance
跨度 span
跨度距离 length of span
跨接电缆 jumper cable
跨接线 jumper
跨路电容 transfer capacitance
跨音速空气动力学 transonic aerodynamics
跨音速压气机 transonic compressor
跨音速叶栅 transonic cascade
快动作触头 quick-action contact
快断开关 quick-break switch
快断熔断器 quick-break fuse
快速接地开关 fault initiating switch
快速励磁 quick response excitation
快速起动灯 instant-start lamp
快速熔断器 fast-acting fuse
快速调节 quick set-up
快速硬化 quick hardening
快速自动重合闸 high-speed automatic reclosing
快中子反应堆 fast neutron reactor
快中子能谱 fast neutron spectrum
快中子通量 fast neutron flux
宽(频)带数据 WBD (wide-band data)
宽(频)带数据链路 WBDL (wide-band data link)
框架 carcass
框图 block diagram
馈电电缆 feeder cable
馈电线 feeder
馈电线断路器 feeder breaker
馈路 feeder
溃散性 collapsibility
醌氢醌半电池 quinhydrone half cell
醌氢醌电极 quinhydrone electrode
扩风器 diffuser
扩散 diffusion
扩散率 diffusivity
扩散器 diffuser
扩压器 diffuser

L

拉杆绝缘子 link insulator
拉杆式定子框架 tie-rod stator frame
拉格朗日法 Lagrangian method
拉格朗日方程 Lagrange('s) equation
拉筋 lacing wire
拉紧机 stretcher
拉紧绝缘子 strain insulator
拉紧螺栓 tie bolt
拉紧装置 strainer
拉莫尔频率 Larmor frequency
拉普拉斯场 Laplace's field
拉普拉斯定律 Laplace's law
拉普拉斯方程 Laplace's equation
拉普拉斯-高斯分布 Laplace-Gauss distribution
拉普拉斯平面 Laplace plane
拉普拉斯算子 Laplace operator
拉入绕组 pull-through winding
拉伸 stretching
拉伸机 stretcher
拉线 stay
拉线V型塔 guyed V tower
拉直机 straightener
拉制生长 growing by pulling
蜡封 wax seal
蜡克盘 lacquer disk
莱顿瓶 Leyden jar
兰姆波 Lamb wave
兰姆移位 Lamb shift
朗伯 lambert
朗道阻尼 Landau damping
朗肯温度 Rankine degree
朗缪尔频率 Langmuir frequency
劳厄法 Laue method
劳伦斯管 Lawrence tube
老化 burn-in
勒 lethargy; lx (lux)
勒布朗克励磁机 Leblanc exciter
勒布朗克联结 Leblanc connection
勒克朗谢电池 Leclanché cell
勒克斯 lx (lux)
勒克斯计 luxmeter
勒秒 lux second
勒纳德射线 Lenard rays
勒谢尔线 Lecher line; Lecher wire
雷电冲击 lightning impulse
雷电过电压 lightning overvoltage
雷电浪涌 lightning surge
雷电流 lightning current
雷电流陡度 lightning current steepness
雷电路径 lightning path
雷管 priming
雷击 lightning stroke
雷诺数 Reynolds number
累积量 cumulant
累加器 accumulator
累接网络 iterative network
累接阻抗 iterative impedance
肋片 fin
类似物 analogue
楞次定律 Lenz's law
冷风系统 quenching system
冷剂水 water as refrigerant
冷凝器 condenser
冷凝器漏水 water leakage in condenser
冷凝物 condensate

冷起动灯 instant-start lamp
冷却剂 coolant
冷却介质 coolant; quenching compound; quenching medium
冷却水 quenching water
冷却液 quenching compound
冷缩配合 shrink fit
冷阴极充气管 tactron
冷油器 oil cooler
冷渣设备 quencher
离地净高 ground clearance
离合器 clutch
离化团粒束 ionized cluster beam
离析 segregation
离心分离 centrifugation
离心机 centrifuge
离心率 eccentricity
离心式压气机 centrifugal compressor
离子 ion
离子导电 ionic conduction
离子阀器件 ionic valve device
离子化 ionization
离子阱 ion trap
离子选择场效应晶体管 ion-selective field effect transistor
离子选择电极 ion-selective electrode
离子烟雾探测器 ionization smoke detector
离子注入 ion implantation
离子注入MOS电路 ion-implanted MOS circuit
李雅普诺夫定理 Liapunov's theorem
李雅普诺夫法 Liapunov's method
李雅普诺夫函数 Liapunov function
理论数值孔径 theoretical numerical aperture
锂电气石 elbaite
力测电流计 electrodynamometer
力矩 moment
力矩电动机 torque motor
力矩器 torquer
力矩旋转执行机构 torque rotary actuator
力学 mechanics
立波 clapotis
立面图 elevation
立式平衡机 vertical balancing machine
励磁 excitation
励磁变压器 excitation transformer
励磁电路 energizing circuit
励磁机 exciter
励磁机响应 exciter response
励磁绕组 excitation winding
励磁设备 energizing apparatus
励磁位置 energized position
励磁响应 excitation response
励弧管 excitron
利用光通量 utilized flux
沥青 bitumen
沥青质体 bituminite
粒子加速器 atom smasher
连杆 coupler
连接 connection; hook-up; interface; junction
连接点 tie point
连接管 tube connector
连接盘 land
连接器 adaptor; connector; interconnection
连接条 intercell connector
连接跳线 tie jumper
连续加感电缆 Krarup cable
连续排放监测系统 CEMS (continuous emission monitoring system)
连续性 continuity
连续自动生产线 transfer machine
连续自激振荡 hunting
联动开关 gang switch
联动控制开关 gang control switch
联动装置 linkage; linkwork
联管节 coupling
联结 connection

联络变压器 tie-in transformer
联络点 tie point
联络线 tie line
联锁插座 interlocked socket outlet
联锁电磁铁 interlocking electromagnet
联锁电路 interlocking circuit
联锁机构 interlocking device
联锁解除 interlock cancelling
联锁开关装置 interlocking switch group
联锁可断开插座 interlocked switched socket outlet
联锁模块 interlocking module
联锁旁通 interlock bypass
联锁旁通开关 interlock bypass switch
联锁跳闸 intertripping
联锁跳闸欠范围保护 intertripping underreach protection
联锁信号 interlocking signal
联锁组合开关 interlocking switch group
联箱 header
联轴节 coupling
联轴器 coupling
链轮 sprocket
链熔线 link fuse
链式电路 LC (link circuit)
链条炉排 travelling grate
链系 linkage; linkwork
良品 non-defective
量程 span
量度继电器 measuring relay
量管 burette
量角器 goniometer
量热器 calorimeter
两段抽汽循环 two-point extraction cycle
两极电机 two-pole machine
两相边界层 two-phase boundary layer
两相变压器 two-phase transformer
两相电路 two-phase circuit
两相发电机 two-phase generator
两相交混 two-phase mixing
两相接地故障 two-phase ground fault
两相流 two-phase flow
两相流动工况 two-phase flow regime
两相流体 two-phase fluid
两相四线制 two-phase four-wire system
两相伺服电机 two-phase servomotor
两相态 two-phase state
两相系统 two-phase system
亮度 brightness
亮度测温法 radiance thermometry
亮度温度 radiance temperature
亮晶 spar
量 magnitude
量化 quantization
量级 magnitude
量子 quantum
列表 tabulation
列表时序控制 tabular sequence control
列氏度 Réaumur scale
邻位化合物 ortho-compounds
临界点 critical point; transformation point; transformation temperature
临界故障分析 failure criticality analysis
临界失效分析 failure criticality analysis
淋水盘 water spraying tray
灵敏电流计 galvanometer
灵敏度 sensibility
灵巧电池 smart battery
零部件清单 PL (parts list)
零长弹簧重力仪 zero-length spring gravimeter
零点 zero (point)
零点残余电压 residual voltage at zero

零点检流计 null galvanometer
零点校准气 zero calibration gas
零点偏移 zero offset
零点漂移 zero drift
零点输出 zero-point output
零点提升 zero elevation
零点误差 zero error
零电感 zero inductance
零电容 zero capacitance
零电压 no-voltage
零固形物处理 AVT (all volatile treatment)
零基线性度 zero-based linearity
零基一致性 zero-based conformity
零排放 zero discharge
零漂移 zero drift
零气 zero gas
零输出 zero output
零输出基线 zero output baseline
零输入响应 zero-input response
零位 isolated position
零位电压 null (position) voltage
零位法 null method
零位调整 null adjustment; zero adjustment
零位位移值 zero displacement value
零相位误差 null phase error
零星批量 isolated lot
零序电流保护 zero-sequence current protection
零序分量 zero-sequence component
零序继电器 zero-phase-sequence relay
零状态响应 zero-state response
流 current; lumen
流程气相色谱仪 process gas chromatograph
流程质谱计 process mass spectrometer
流化床燃烧 fluidized-bed combustion
流化床涂敷 fluidized-bed coating
流颈 vena contracta
流明 lumen
流水作业 tact system
流速仪 current meter
流线 streamline
硫 sulphur
硫化 sulphurization
硫化铁矿类 pyrites
硫化物废水 wastewater sulphide
硫酸铝 alum
硫酸铜 copper sulphate
六相电路 hexa-phase circuit
漏 leak
漏斑 holiday
漏磁通 leakage flux
漏地电流 earth leakage current
漏地电流继电器 earth leakage relay
漏地探测 ground leakage detection
漏点 holiday
漏电 leak; leakage
漏电继电器 leakage relay
漏电率 leakage rate
漏电阻 ohmic leakage
漏感 leakage inductance
漏气试验 gas leakage test
漏汽损失 leakage loss
漏损 ullage
漏泄场 leakage field
漏泄电导 leakance
漏泄电流 leakage current
漏泄电阻 leak
漏泄光波导 leaky light guide
漏泄检验器 leak detector
漏泄损失 leakage loss
漏泄系数 leakage coefficient
漏液 leakage
卢瑟福背散射谱法 RBS (Rutherford backscattering spectroscopy)
炉胆 furnace
炉料 charge
炉前设备 BFE (boiler-front equipment)
炉身 stack
炉膛 furnace

炉膛防爆保护 furnace explosion protection
炉膛内爆 furnace implosion
炉膛设计压力 furnace enclosure design pressure
炉渣 slag
卤水 brine
鲁棒控制 robust control
鲁棒控制器 robust controller
鲁棒性 robustness
录波器 oscillograph; oscilloscope
路径 path
伦纳德控制 Leonard control
轮毂 hub
轮毂比 tip-hub ratio
轮周效率 wheel efficiency
螺杆式压缩机 screw compressor
螺孔钻 tap drill
螺口灯头 screw base; screw cap
螺口灯座 Edison socket
螺帽 nut
螺帽形开关 nut switch
螺母 nut
螺栓 bolt
螺丝底孔钻 tap drill
螺纹接头 nipple
螺纹中径 pitch diameter
螺旋铂灯丝 platinum spiral filament
螺旋灯座 Edison base
螺旋管 serpentine tube
螺旋千斤顶 jack screw
螺旋千斤顶系统 jack-screw system
螺旋式灯头 Edison socket
螺旋输送式炉 screw conveyor furnace
螺旋形电阻体 spiral resistor
螺旋钻 auger
裸丝敏感元件传感器 sensor with bare wire sensitive element
裸线 naked wire
落地雷 ground flash
落地雷密度 ground flash density
铝 aluminium
铝青铜 aluminium bronze
铝土矿 bauxite
履带式车辆 crawler
氯 chlorine
氯胺 chloramine
氯化 chlorination
氯化铵 ammonium chloride
氯化聚氯乙烯 CPVC (chlorinated polyvinyl chloride)
氯化聚醚 CPE (chlorinated polyether)
氯化器 chlorinator
氯化物 chloride
氯离子 chloride ion
滤波电容器 filter capacitor
滤波器 (wave) filter
滤水 water filtration
滤网 strainer
滤芯 filter cartridge

马尔可夫过程 Markov process
马尔可夫链 Markov chain
马赫角 Mach angle
马赫数 Mach number
马赫效应 Mach effect
马赫锥 Mach cone
马克斯发生器 Marx generator
马力 horsepower
马蹄形磁铁 horseshoe magnet
马希森标准 Matthiessen's standard
码 code
埋弧加热 submerged arc heating
埋弧炉 submerged arc furnace
麦克劳林展开 Maclaurin expansion
麦克斯韦 Mx（maxwell）
麦克斯韦常数 Maxwell constant
麦克斯韦电桥 Maxwell bridge
麦克斯韦定则 Maxwell rule
麦克斯韦方程组 Maxwell equations
麦克斯韦模型 Maxwell model
脉搏传感器 pulse sensor; pulse transducer
脉冲编码调制 PCM（pulse code modulation）
脉冲波形 impulse shape
脉冲持续时间 pulse duration
脉冲重复频率 PRF（pulse repetition frequency）
脉冲电流 impulse current
脉冲电容 impulse capacitance
脉冲堆积 pulse pile-up
脉冲发生器 impulse generator
脉冲负荷 impulse load
脉冲傅里叶变换核磁共振波谱仪 pulsed Fourier transform NMR spectrometer
脉冲傅立叶变换核磁共振法 pulsed Fourier transform NMR
脉冲函数 impulse function
脉冲回波法 pulse echo method
脉冲回转角 pulse flip angle
脉冲击穿 impulse breakdown; impulse flashover
脉冲击穿强度 impulse breakdown strength
脉冲击穿试验 impulse flashover test; impulse sparkover test
脉冲极谱仪 pulse polarograph
脉冲计数器 impulse meter
脉冲交流电 impulse alternating current
脉冲控制 pulse control
脉冲宽度 pulse duration; pulse width
脉冲宽度调制 pulse duration modulation
脉冲频率 impulse frequency
脉冲式航空电磁仪 pulse-type airborne electromagnetic instrument
脉冲输入 pulse input
脉冲调频控制系统 pulse frequency modulation control system
脉冲位置调制 pulse position modulation
脉冲噪声 impulse noise
脉冲振幅分析 kicksort
脉冲振幅分析器 kicksorter
脉动 pulsation

脉动水银管　impulsing mercury tube
脉宽调制　pulse duration modulation
脉宽调制逆变器　pulse-width modulated inverter
脉码调制　PCM (pulse code modulation)
脉泽　maser
满容量分接　full-power tapping
满载　full load
曼尼希碱　Mannich base
漫射　diffusion
毛巾架　towel rack
铆钉　rivet
铆接　rivet
帽槽　clevis
帽窝　socket
煤焦油　coal tar
煤沥青　coal tar
煤泥　slime
煤烟　soot
每转走刀量　feed per revolution
门　gate
门电路　gate circuit
门极　gate electrode
门极电压　gated voltage
门极控制　gate control
门信号　gate signal
蒙特卡罗法　Monte Carlo method
米　m (metre)
米勒代码　Miller code
米勒时基　Miller time base
米勒效应　Miller effect
米勒振荡器　Miller oscillator
密闭隔离挡板　tight-closing isolating damper
密度　density
密度计　densimeter
密封　sealing
密封检查　leakage check
密封圈　gasket
密封套　gland
密封型电机　impervious machine
密封油　sealing oil

密排管结构　tangent tube construction
密排管水冷壁　tangent tube wall
密实度　packing
密锁自动耦合器　tight-lock coupler
密钥生成器　key generator
面板　panel
面积仪　planimeter
面缩率　reduction of area
面向过程仿真　process-oriented simulation
面罩　face shield
灭磁　excitation suppression
灭磁绕组　killer winding
灭弧点　quenching moment
灭弧电路　quench circuit
灭弧电阻　quenching resistance
灭弧时间　quenching time
灭弧时刻　quenching moment
灭弧室　interrupt chamber; interrupter
敏感度　susceptibility; susceptivity
敏感器　sensor
敏感元件　sensor
明矾　alum
明线线路　open-wire circuit
模块　module
模量　modulus
模拟　simulation
模拟电路　analogue circuit
模拟应变　simulating strain
模拟应变装置　simulating strain device
模数　module; modulus
模数转换器　ADC (analogue-digital converter); analogue-to-digital converter
膜片　diaphragm
膜(态)沸腾　film boiling
摩(尔)　mol (mole)
磨耗　abrasion
磨口接头　ground joint
磨煤废气　waste gas of coal pulverization

磨蚀度 abrasivity
磨损 abrasion; attrition
末级加热器 top heater
母板 motherboard
母版 master mask
母片 master slice
母线 busbar; main lead
姆欧 mho
姆欧表 mhometer
目标 target

目标变量 target variable
目标捕获 target acquisition
目标程序库 object program library
目标辐照 target irradiation
目标函数 objective function
目标模块 object module
目标区 target area
目标燃耗 target burn-up

纳 noy
纳安 na (nanoampere)
纳安计 nanoammeter
纳安培 na (nanoampere)
纳法 nanofarad
纳亨 nanohenry
纳米 millimicron; nm (nanometre)
纳米复合材料 nanocomposite
纳米级电路 nanocircuit
纳秒 ns (nanosecond)
纳瓦 nanowatt
奈奎斯特定理 Nyquist's theorem
奈奎斯特-柯西判据 Nyquist-Cauchy criterion
奈奎斯特频率 Nyquist frequency
奈奎斯特曲线 Nyquist curve
奈培 napier; Np (neper)
奈塞 nesa
耐短路变压器 short-circuit-proof transformer
耐故障能力 fault withstandability
耐光性 light resistance
耐久性 durability
耐漏液的 leakproof
耐热概貌 thermal endurance profile
耐热试验 heat run
耐压试验 high-voltage test
耐用性 durability
挠性表面加热器 flexible surface heater
挠性接头 kidney joint
内禀磁性 intrinsic magnetic property
内部尺寸 inside dimension

内部等效电压 internal equivalent voltage
内部电弧故障 internal arcing fault
内部电容 internal capacitance
内部对讲电路 talk-back circuit
内部放电 internal discharge
内部故障 internal fault
内部故障测试 internal fault test
内部观测系统 internal viewing system
内部过压 internal overvoltage
内部过压断路器 internal overpressure disconnector
内部抗干扰性 internal immunity
内部冷却导体 inner-cooled conductor
内部冷却导线 inner-cooled conductor
内部轮廓 inside contour
内部漂移场 internal drift field
内部缺陷 in-zone fault
内部剩余电压 internal residual voltage
内部释放 internal discharge
内存储信息位置图示 topogram
内电渗 electroendosmosis
内对流敏感元件传感器 sensor with internal convection sensitive element
内光电效应 internal photoelectric effect
内过电压 internal overvoltage
内建电场 internal electric field
内接螺母 female nipple
内接线 internal wiring
内径 inside diameter

内径测量仪 inside calipers
内聚力 cohesion
内聚性 cohesion
内绝缘 internal insulation
内卡钳 inside calipers
内孔连接器 female connector
内宽 inside width
内螺纹 internal thread
内螺纹连接器 female connector
内燃机 internal combustion engine
内务处理数据 housekeeping data
内效率 internal efficiency
内置仿真器 in-circuit emulator
内置时钟 internal clock
内轴承环 inner bearing ring
内装轴承 inboard bearing
内装转子 inboard rotor
内阻 internal resistance
能见度 visibility
能见度表 visibility meter
能见度目标物 visibility marker; visibility object
能见范围 visual range
能量学 energetics
能谱法 spectroscopy
能容 capacity
能斯特-爱因斯坦关系 Nernst-Einstein relation
能斯特灯 Nernst lamp
能斯特电桥 Nernst bridge
能斯特探测器 Nernst detector
能斯特效应 Nernst effect
尼科尔斯图 Nichols chart
尼特 nit
泥浆 slurry
泥渣 sludge
拟合 fitting
逆变换 inverse transformation
逆变器 inverter
逆变器触发装置 inverter trigger set
逆变器稳定限度 inverter stability limit
逆变器终止控制 inverter termination control
逆变效率 inversion efficiency
逆变因数 inversion factor
逆并联 inverse parallel connection
逆电流 inverse current
逆弧 arc-back
逆时分级整定 inverse-time grading
逆时进给率 inverse-time feed rate
逆时延时 inverse-time lag
逆时针 CCW (counterclockwise)
逆转特性继电器 inverse-characteristic relay
年负荷率 yearly load factor
年负荷曲线 yearly load curve
年负载率 yearly load factor
年负载曲线 yearly load curve
年平均雷电日水平 isokeraunic level
年效率 yearly efficiency
年终调整 year-end adjustment
年最高负荷 yearly maximum load
年最高负载 yearly maximum load
黏度 viscosity
黏度传感器 viscosity sensor; viscosity transducer
黏度计 viscometer
黏度天平 viscosity balance
黏附 adhesion
黏垢 slime
黏合 adhesion
黏合剂 adhesive
辗轮混砂机 muller
碾压机 mangle
啮合 engagement
镍电极 nickel electrode
镍铁电池 Edison battery; NiFe cell
镍铁合金 NiFe
镍铁蓄电池 NiFe accumulator
凝固 coagulation; solidification
凝结水 condensate

凝结水泵 CP (condensate pump)
凝聚 coagulation
凝聚剂 agglomerant
凝汽器 condenser
凝水泵 CP (condensate pump)
牛顿迭代法 Newton iteration method
牛顿定律 Newton's law
牛顿-拉弗森法 Newton-Raphson method
牛皮纸 kraft (paper)
扭摆 torsional pendulum
扭绞系数 lay ratio
扭矩 twisting moment
扭力 torsion
扭力计 torsion meter
扭力阻尼器 torsional damper
扭量 twistor
扭叶片 twisted blade
扭应力 twisting stress
扭折 kink
扭转 torsion
扭转角 torsion angle
扭转力矩 twisting moment
扭转临界转速 torsional critical speed
扭转振荡 torsional oscillation
扭转轴 torque shaft
浓缩物 concentrate
努普硬度压头 Knoop hardness penetrator
努普硬度值 Knoop hardness number
诺顿变换 Norton transformation
诺顿定理 Norton's theorem
诺模图 nomogram; nomograph

欧 ohm
欧安表 ohm ammeter
欧姆 ohm
欧姆电桥 ohmic bridge
欧姆电阻 ohmage；ohmic resistance
欧姆定律 Ohm's law
欧姆计 ohmmeter
欧姆接触 ohmic contact
欧姆接合 ohmic junction
欧姆损失 ohmic loss
欧姆阻抗 ohmage
偶极子双极子 dipole
偶然故障 incidental defect
偶然性 contingency
耦合 coupling
耦合器 coupler

爬电距离 creepage distance
爬行放电 creepage discharge
帕(斯卡) Pa (pascal)
排出 blowdown
排出压头 discharge head
排除废气 scavenge
排队存取法 queued access method
排队控制算法 queue algorithm
排队论 queueing theory
排队顺序存取法 queued sequential access method
排放 discharge
排放装置 discharger
排废 waste discharge
排挤厚度 squeezing thickness
排架 bent
排量 displacement
排气孔 vent hole
排水 drainage
排水工程 sewerage; wastewater engineering
排水管 drainage pipe; water-discharge tube
排水管渠 sewer
排水量 water discharge
排污阀 waste valve
盘管 coil
盘式装料装置 pan charger
盘旋场 nutation field
盘旋馈电 nutating feed
抛物线 parabola
泡沫共腾 foaming
佩尔捷效应 Peltier effect
配电板 keyset
配电变压器 distribution transformer
配电管理系统 DMS (distribution management system)
配电间 incoming cubicle
配电盘 switchboard
配电装置 switchgear
配电自动化 distribution automation
配合公差 fit tolerance
配水器 water divider
配置 configuration
喷管 nozzle
喷漆薄膜 lacquer film
喷枪 lance
喷燃器 burner
喷射泵 ejector
喷射阀 injection valve
喷射管 adjutage
喷射混凝土 gunite; shotcrete
喷射器 kicker
喷射压缩机 jet compressor
喷水泵 injection pump; water-jet pump
喷水床 water spouted bed
喷水阀 water spray valve
喷水冷凝器 water-jet condenser
喷水冷却 water spray cooling
喷水枪 water lance
喷雾器 sprayer
喷油泵 injection pump
喷嘴 nozzle
喷嘴导叶 nozzle blade
喷嘴式流量计 nozzle flowmeter
喷嘴系数 nozzle coefficient
硼 boron
膨胀 bellying; dilatation
膨胀性 dilatability
碰垫 fender

批量供应　bulk supply
皮带运输机　belt conveyor
皮托管　Pitot tube
皮重　tare (weight)
疲劳　fatigue
疲劳断裂　fatigue fracture
疲劳断裂试验　fatigue fracture test
疲劳破裂　fatigue failure
疲劳失效　fatigue failure
匹配变压器　matching transformer
匹配波导管　matched waveguide
匹配电路　matching circuit
匹配连接　matched junction
匹配滤波器　matched filter
匹配筛选器　matched filter
匹配衰减器　matching attenuator
偏离　bias
偏频　offset frequency
偏心距　eccentricity
偏心轮机构　eccentric mechanism
偏心受压　eccentric compression
偏心位置　eccentric position
偏心压缩　eccentric compression
偏心载荷　eccentric load
偏压　bias
偏移　offset
偏移电流　offset current
偏移电压　offset voltage
偏移量　offset
偏移频率　offset frequency
偏移射束　offset beam
偏置　bias; offset
偏转板　deflector
片电阻　sheet resistor
片剂　tablet
片间绝缘　lamination insulation
片晶　lamella
漂移　drift
漂白剂　bleach
撇渣　skim
频带　band
频率计　cymometer
频率继电器　frequency relay

频敏变阻器　frequency-sensitive rheostat
频谱　spectrum
频谱分析仪　spectrum analyzer
频谱密度　spectral density
频域　frequency domain
品质管理体系　QMS (quality management system)
品质环　quality loop
品质指标　QI (quality index)
平行度　parallelism
平行逆变器　parallel inverter
平衡电抗器　interphase reactor
平衡电路　equalization circuit
平均电流　mean current
平均功率　mean power
平均可用度　mean availability
平均偏差　mean deviation
平均频率　mean frequency
平面　plane
平面布置图　layout plan
平面敷设　flat formation
平面图　plan
平头钉　tack
平头焊　tack welding
平移波　wave of translation
屏蔽　screen
屏蔽绝缘电缆系统　insulated-shield cable system
屏蔽容器　coffin
屏蔽双线馈线　twinax
屏蔽效能　screening effectiveness
破坏负载　failure load
破坏荷载　failure load
破坏压力　rupture pressure
铺设　installation
普朗特数　Prandtl number
谱　spectrum
谱带宽度　spectral bandwidth
谱分解　spectral resolution
谱库检索　library searching
谱密度　spectral density
谱线　spectral line
曝气　aeration
曝气器　aerator

漆 lacquer
漆包绝缘 enamel insulation
漆包线漆 wire enamel
齐纳击穿 Zener breakdown
启闭机 hoist
启动 start-up
启动分配 initiation assignment
启动前检查 pre-start checking
启用 invocation
起电盘 electrophore; electrophorus
起拱 camber
起痕蚀损 tracking erosion
起始功率 initial watts
起始浪涌电压分布 initial surge-voltage distribution
起始瞬态电抗降 initial transient reactance drop
起始瞬态恢复电压 ITRV (initial transient recovery voltage)
起始应力 initial stress
起重滑车 hoisting tackle
起重机 crab; hoist; lifter
起重螺栓 jack bolt
起重螺旋 jack screw
起重装置 hoisting gear; jacking device
气涤器 scrubber
气动泵 pneumatic pump
气动厚度计 pneumatic thickness meter
气动活塞式压力计 pneumatic piston gauge
气动极限操作器 pneumatic limit operator
气动控制 pneumatic control
气动跑兔 pneumatic rabbit
气动系统 pneumatic system
气动执行机构 pneumatic actuator
气锅 steamer
气焊 gas welding
气孔 pore
气密性试验 gas leakage test
气密与汽密型电机 gas and vapour proof machine
气泡 blister
气溶胶 aerosol
气蚀 cavitation
气体电离过程 electromerism
气体电离检定法 electroscopy
气体绝缘电路 gas-insulated circuit
气体绝缘套管 gas-insulated bushing
气雾剂 aerosol
气隙 air gap
气旋 cyclone
气穴 cavitation
气压传感器 baroceptor
气压计 air gauge; barometer
气压式调压室 pneumatic surge chamber
气压温度计 barothermograph
气压系统 pneumatic system
气压扬水机 aqua thruster
气闸 airlock
汽包 drum
汽包就位 placement of boiler drum
汽动锅炉给水泵 TBFP (turbo boiler feed pump)
汽缸 cylinder
汽锅 kettle

汽化 vaporization
汽化器 carburettor
汽轮发电机 turbo-generator
汽轮机 steam turbine
汽轮机超速保护 turbine over-speed protection
汽轮机旁路 TBP (turbine bypass)
汽轮机旁路控制 TBC (turbine bypass control)
汽轮机损失 turbine loss
汽轮机-压气机组 TC (turbine-compressor)
汽轮机振动保护 turbine vibration protection
汽轮机自动控制 ATC (automatic turbine control)
汽轮给水泵 turbine-driven feed pump
汽室 steam chest
汽水分离装置 water separator
汽水共腾 priming
汽转球 aeolipile
千安(培) KA (kiloampere)
千奥(斯特) kilo-oersted
千巴 kbar (kilobar)
千磅 kip
千磁力线 kiloline
千达因 kilodyne
千电子伏特 keV (kilo-electronvolt)
千电子伏级加速器 kevatron
千乏 kvar (kilovar)
千乏时 kilovar-hour
千分率 permillage
千伏 kV (kilovolt)
千伏安 KVA (kilovolt-ampere)
千伏电压 kilovoltage
千伏特 kV (kilovolt)
千高斯 KG (kilogauss)
千赫(兹) kHz (kilohertz)
千焦(耳) kJ (kilojoule)
千斤顶 jack
千居里 kilocurie
千卡 kcal (kilocalorie)
千克 kg (kilogram)
千勒(克斯) kilolux
千流明 kilolumen
千伦琴 kiloroentgen
千升 kl (kilolitre)
千瓦 kW (kilowatt)
千瓦安 kWa (kilowatt-ampere)
千瓦时 kWh (kilowatt-hour)
千微升 kilolambda
千位 kilobit
千兆 kilomega
千兆位 kilomegabit
千兆周 kilomegacycle
千周 kc (kilocycle)
千周波电磁法 KEM (kilocycle electromagnetics)
千字节 Kb (kilobyte)
钎焊 soldering
牵入转距 pull-in torque
牵引板 towing plate
牵引电动机 traction motor
牵引电流 traction current
牵引发电机 traction generator
牵引负载 traction load
牵引式发动机 traction engine
牵引式锅炉 traction boiler
牵引线 fishing wire
铅 lead
铅包电缆 lead clad cable; lead covered cable
铅耗 lead loss
铅皮 lead covering
铅屏 lead shield
前陡度 wavefront steepness
前进相位 travelling phase
前缘 leading edge
前置放大器 preamplifier
前置(式)汽轮机 top(ping) turbine
钳 nippers
钳工 fitter
钳式电流表 tong-test ammeter
钳形电流表 hook-on ammeter
潜水电线 water immersion wire
潜水型电动机 water-submerged motor
欠电流继电器 under-current relay

欠电压保护 under-voltage protection
欠范围式纵联保护 under-reaching pilot protection
欠载 underload
强电工程 heavy-current engineering; heavy electrical engineering
强电流 heavy current
强度 intensity; strength
强度极限 ultimate strength
强耦合 tight coupling
强迫冷却 forced cooling
强相 leading phase
强制实施 mandatory implementation
墙角圆 bead
墙式过热器 wall superheater
桥吊 bridge crane
桥接电缆 jumper cable
桥接配置 jumper assignment
桥式起重机 bridge crane
桥台 abutment
翘板型轮毂 teetered hub
翘曲 warp
切断比 interruptive ratio
切断电源 isolation from supply
切割点铣刀半径轨迹校正 intersection cutter radius compensation
切换 switch
切换触点 transfer contact
切换阀 transfer valve
切换继电器 transfer relay
切换母线 transfer bus
切换片 transfer strip
切击式水轮机 tangential (flow) turbine
切面 tangent plane
切深进给运动 infeed motion
切碎机 shredder
切线 tangential line; tangent (line)
切线逼近法 tangential approximation method
切线刚度矩阵 tangent stiffness matrix
切向波 tangential wave
切向非整周进水式水轮机 tangential partial turbine
切向分量 tangential component
切向加速度 tangential acceleration
切向剪应力 tangential shearing stress
切向键槽 tangential keyway
切向接管 tangential nozzle
切向进汽 tangential admission
切向力 tangential force
切向力偶 tangential couple
切向力系数 tangential force coefficient
切向燃烧 tangential firing
切向燃烧锅炉 tangentially fired boiler
切向燃烧炉膛 tangentially fired combustion chamber; tangentially fired furnace
切向束射管 tangential beam tube
切向四角燃烧 tangential corner firing
切向速度 tangential velocity
切向梯度探头 tangential gradient probe
切向推力 tangential thrust
切向叶片间距 tangential blade spacing
切向异相位 tangential out-of-phase
切向应变 tangential strain
切向应力 tangential stress
轻子 lepton
氢冷电机 hydrogen-cooled machine
倾倒式炉 tilting furnace
倾覆力矩 tilting moment
倾炉架 tilting cradle
倾斜 bevel; obliquity
倾斜淀积 oblique deposition
倾斜式微压计 tilting micromanometer
倾斜压力计 tilting manometer

倾斜阳极 oblique anode
倾斜仪 clinometer
倾卸装置 tilter
清澈度 limpidity
清除 scavenge
清除剂 scavenger
清管器 scraper
清绘 fair drawing
清基 foundation cleaning
清漆 lacquer; varnish
清晰度 articulation
擎住电流 latching current
求积仪 planimeter
球阀 BV (ball valve)
球形度 sphericity
区间 interval
区内故障 in-zone fault
区熔生长 growing by zone melting
曲率 curvature
曲线 curve
曲折形联结 zigzag connection
曲轴 crankshaft
驱动器 actuator; driver
屈服点 yield point
屈服强度 yield strength
屈服应力 yield stress
屈服准则 yield criterion
趋电性 electrotaxis; electrotropism
趋肤效应 Kelvin effect
取样辐射仪 sampling radiometer
取样时间 sampling time
去湿器 dehumidifier
去污 decontamination
去谐滤波器 harmonic filter
全波结构 full-wave arrangement
全电流 total current
全电容 total capacitance

全堆芯事故 WCA (whole core accident)
全方位指示器 omni-bearing indicator
全辐射高温计 total radiation pyrometer
全辐射通量计 total radiation fluxmeter
全固形物 total solids
全国超高压电网 national super grid
全挥发性处理 AVT (all volatile treatment)
全交换容量 total exchange capacity
全控联结 fully controllable connection
全能加速器 omnitron
全气孔率 true porosity
全身计数器 WBC (whole-body counter)
全甩负荷 total load rejection
全水分 total moisture
全脱氧钢 killed steel
全微分 total differential
全向辐射 omnidirectional radiation
全向接收 omnidirectional reception
全向天线 omnidirectional antenna
全谐波畸变 total harmonic distortion
全振幅 total amplitude
缺口冲击试验 impact notch test
缺陷 blemish
群时延 group delay

燃点 fire point; kindling point
燃耗 burn-up
燃料电池 fuel cell
燃料电池组 fuel battery
燃料计量 fuel measurement
燃料配用 fuel blending
燃料芯块 fuel pellet
燃料元件〖带横向肋的〗 transversely-finned fuel element
燃料肿胀 fuel swelling
燃料组件间水隙 water gap between fuel assemblies
燃气轮(发电)机组 gas turbine set
燃烧 combustion
燃烧器 burner; combustor
燃烧室 combustor
燃烧脱硫 desulphurization during combustion
燃烧自动调节 ACC (automatic combustion control)
扰码器 scrambler
绕包绝缘 lapped insulation
绕包线 lapped wire
绕线架 drum stand
绕组 coil; winding
绕组间故障 interwinding fault
绕组节距 winding pitch
绕组线段短路 inter-stand short circuit
绕组线圈 winding coil
绕组因数 winding factor
热备用 hot standby
热崩溃 thermal runaway
热泵 heat pump
热电放射效应 Edison effect
热电联产 cogeneration

热电偶 TC (thermal couple; thermocouple)
热-电牵引 thermo-electric traction
热电势 thermoelectric potential
热辐射体 thermal radiator
热固性塑料 thermoset plastics
热挂 thermal blockage
热耗量 heat consumption
热虹吸器 thermosiphon
热汇 heat sink
热降额因数 thermal derating factor
热控制系统 TCS (thermal control system)
热扩散率 thermal diffusivity
热离子发射 thermionic emission
热离子管 valve
热力学系统 thermodynamic system
热力学循环 thermodynamic cycle
热量 quantity of heat
热量计 calorimeter
热量平衡 TB (thermal balance)
热流通量 heat flux
热敏电阻(器) thermistor
热膨胀系数 TEC (thermal expansion coefficient)
热疲劳 thermal fatigue
热偏差 thermal deviation
热平衡 heat balance; TB (thermal balance)
热容 heat capacity; thermal capacitance
热容量 heat capacity
热失控 thermal runaway

热失重法 thermogravimetry
热寿命 thermal life
热双金属 thermo-bimetal
热损失 thermal loss
热态启动 hot start-up
热套配合 shrink fit
热稳定试验 thermal stability test
热析 sweating
热线 hot wire
热效率 TE (thermal efficiency)
热压铸 injection moulding
热延时开关 thermal time-delay switch
热影响区 HAZ (heat affected zone)
热源 heat source
热运行 heat run
热再启动 hot restart
热噪声 Johnson noise
热中子 thermal neutron
热阻 thermal resistance
人工重调 hand reset
人工泄水道 sluice
人机接口 human-machine interface
人孔 manhole
人为差错 human error
人为误差 human error
人因工程学 human factors engineering
人字起重机 A-derrick
刃形指针 knife edge pointer
任务分派程序 task dispatcher
任务控制块 TCB (task control block)
任务描述符 task descriptor
任务执行存储器 task execution memory
任选设备 optional equipment
韧铜 tough copper
韧性金属 tough metal
韧性试验 toughness test
容错 fault tolerance
容积法 volumetric method
容积式流量计 positive displacement flowmeter; volumetric flowmeter
容积式水表 volumetric water meter
容量 capacity
溶剂 solvent
溶解 dissolution
溶解度 solubility
溶解物 solute
溶解氧 dissolved oxygen
溶液 solution
溶质 solute
熔断短路电流 fused short-circuit current
熔断器 fuse
熔断体 fuse-link
熔管 cartridge
熔剂 flux
熔件 fuse element
熔炉 smelter
熔丝开关 switch-fuse
熔线片 link fuse
熔渣 slag
冗余单元 redundancy unit
冗余计算机系统 redundancy computer system
冗余设备 redundancy unit
冗余信息 redundant information
柔性石墨 flexible graphite
蠕变极限 ultimate creep
乳化剂 emulsifier
乳化器 emulsifier
乳胶板 emulsion plate
乳液 emulsion
入口 inlet; input
入侵报警系统 intruder alarm system
软磁材料 soft magnetic material
软化 softening
软化剂 softener
软化器 softener
软联结 flexible connection
软木塞 cork
软水处理装置 water softening plant
软线 cord

润滑剂 lubricant
润滑器 lubricator
润滑油 lubricant

弱场 feeble field
弱电流 light current
弱相 lagging phase

Ss

塞尺 feeler
塞止接头 stop joint
赛璐珞 celluloid
三次风 tertiary air
三次绕组 tertiary winding
三极管枪 triode gun
三极振荡管 oscillion
三相电路 three-phase circuit
三相三柱式铁芯 three-phase three-limb core
三芯分支盒 trifurcator
三芯分支接头 trifurcating joint
三氧化二铝 aluminium oxide
散度 divergence
散射 scattering
散射辐射 scattered radiation
散射光浊度计 scattering turbidimeter
散射计 scatterometer
散射离子能量 scattering ion energy
散射离子能量比值 scattering ion energy ratio
散射体 scatterer
散发火花 spark emission
散热件 heat sink
散热片 fin
扫描 scanning
扫描X射线光谱仪 sequential X-ray spectrometer
扫描电子显微镜 SEM (scanning electron microscope)
扫描辐射温度计 scanning radiation thermometer
扫描离子微区探针 scanning ion microprobe
扫描器 scanner
扫描速率 scan rate
扫描隧道显微镜 STM (scanning tunnelling microscope)
扫描透射电子显微镜 STEM (scanning transmission electron microscope)
色度 colority
色度测量 colorimetry
色度计 colorimeter
色度学 colorimetry
刹车 brake
砂床 bed
砂眼 blister
筛 sieve
筛粉机 bolter
筛子 sieve
栅格输入电压 grid input voltage
栅格输入功率 grid input power
栅格因素 grid factor
栅极 gate electrode; grid
栅极电流 grid current
栅极端子 gate terminal
栅极激励电压 grid driving voltage
栅极激励功率 grid driving power
栅极控制 gate control
栅极脉冲 grid pulse
栅极引出线 gate terminal
栅距 lattice spacing
栅控弧光放电管 grid-controlled arc discharge tube
栅偏压 grid bias
闪电 lightning
闪络试验波尾电压 tail-of-wave impulse test voltage
闪烁室式射气仪 scintillation emanometer

闪烁探测器 scintillation detector
闪烁体 scintillator
扇出 fan-out
扇段 sector
扇区 sector
扇形齿轮 tooth sector
扇形叶片 sector
扇型继电器 vane-type relay
商 quotient
商继电器 quotient relay
商值表 quotient meter
上层线棒 top bar
上池 upper reservoir
上法兰 top flange
上料 loading
上升率抑制器 rate-of-rise suppressor
上鼠笼条 top bar
上水库 upper reservoir
上位计算机 higher-level computer
上位控制回路 higher-level control loop
上位系统 higher-level system
上位自动化系统 higher-level automation system
上弦 top chord
上悬窗 top-hinged window
烧蚀的 ablative
少数载流电流 minority carrier current
少数载流子 minority carrier
少数载流子寿命 minority carrier lifetime
舌状花样 tongue pattern
蛇形管 serpentine tube
设备 apparatus; installation
设备容量 installed capacity
设备最高电压 highest voltage for equipment
设定点 set point
设定使用条件 intended conditions of use
设定寿命 intended life
设计工况 design condition

设计水头 design head
射波器 wave launcher
射极跟随器 emitter follower
射极输出器 emitter follower
射频电极 radio frequency electrode
射频电缆 radio frequency cable
射频敏感器 radio frequency sensor
射汽抽气器 steam-jet air ejector
射水抽气器 water-jet air ejector
射水抽水器 water ejector
射线剂量传感器 radiation dose sensor; radiation dose transducer
摄谱仪 spectrograph
伸臂 cantilever
砷化镓二极管 gallium-arsenide diode
深槽 kettle
甚长波 myriametric wave
甚低频电磁仪 VLF electromagnetic receiver
甚高频全向信标 VOR (VHF omnidirectional range)
渗出液 transudate
渗滤咖啡壶 percolator
渗透 osmosis
渗透管 osmosis tube; permeability tube
渗透率 permeability
渗锌 sheradizing
升 l (litre)
升船机 ship lift
升华 sublimation
升降机 hoist; jacking device; lifter
升降装置 jacking device
升压变压器 step-up transformer; transformer booster
升压机 booster
生物环 biocycle
生物降解 biodegradation
声波定位仪 sonar

声呐 sonar
声呐定位法 sonar
声频电路 audio-circuit
声速 sonic speed
声压级 SPL (sound pressure level)
绳环 becket
剩磁 residual magnetism
剩磁法 residual method
剩磁检查仪 residual magnetic field detector
剩余不平衡 residual unbalance
剩余电流 residual current
剩余电流互感器 residual current transformer
剩余电压 residual voltage
剩余电压互感器 residual voltage transformer
剩余电压绕组 residual voltage winding
剩余浮力 residual buoyancy
失步 out of step; out of synchronism
失步保护 out-of-step protection
失步继电器 out-of-step relay
失步跳闸 out-of-step tripping
失触发 triggering failure
失调 offset
失调电流 offset current
失调电压 offset voltage
失通 firing failure
失相 out of phase
失效保护安全联锁 fail-safe interlock
失效分析 failure analysis
失效概率分布 failure probability distribution
失效率 failure rate
失效模式 failure mode
失效原因 failure cause
失泽物 tarnish
施特鲁哈尔数 Strouhal number
施主电离能 ionizing energy of donor
湿井干井冷却系 WDCS (wet-well dry-well cooling system)
湿球温度计 wet bulb thermometer
湿陷性 collapsibility
十进制 decimal system
十六烷 cetane
十亿位 kilomegabit
十字接头 Hooke's coupling
十字接线盒 intersection box
十字头 cross head
石灰 lime
石墨化碳 graphitized carbon
石英 quartz
石英温度计 quartz thermometer
石油化学品 petrochemical
石油精炼 petroleum extraction
石油探测浮标系统 petroleum oil detection buoy system
石油制品 petroleum
时变参数 time-varying parameter
时变场 time-varying field
时变反应性 time-varying reactivity
时变率 time rate
时变系数 time-varying coefficient
时变系统 time-varying system
时标因子 time-scale factor
时程分析法 time history analysis method
时程曲线 travel-time curve
时分 time division
时分多路乘法器 time-division multiplier
时分复用 TDM (time-division multiplexing)
时分交换 time-division switching
时分开关 TDS (time-division switch)
时分制 time-division system
时基 time base
时基电路 time-base circuit
时基发生器 time-base generator
时基误差 TBE (time-base error)
时间比例 time scale

时间边缘效应 time edge effect
时间变量 time variable
时间变率 time rate
时间标度 time scale
时间标记 time tagging
时间补偿 time bias
时间常数 time constant
时间-沉降曲线 time-settlement curve
时间导数 time derivative
时间电流阈值 time-current threshold
时间反符合电路 time-anticoincidence circuit
时间符合电路 time-coincidence circuit
时间-幅度变换器 TAC (time-to-amplitude converter)
时间固结曲线 time-consolidation curve
时间关联 time correlation
时间恒定调整器 time-invariant regulator
时间恒定系统 time-invariant system
时间划分 time division
时间积分 time integral
时间记录器 time recorder
时间继电器 time relay
时间监视器 time monitor
时间间隔指示器 time-interval indicator
时间距离继电器 time distance relay
时间开关 time switch
时间量程 time scale
时间码生成器 time-code generator
时间脉冲 time pulse
时间脉冲分配器 time pulse distributor
时间片 time slice
时间-频率对偶性 time-frequency duality
时间平均干涉测量术 time-average interferometry
时间平均全息术 time-average holography
时间平均全息图 time-average hologram
时间平均条纹图样 time-average fringe pattern
时间-数字变换器 time-to-number converter
时间锁定 time lock
时间调制 time modulation
时间同步 time lock
时间相干性 time coherence
时间相关 time dependence
时间相角 time-phase angle
时间相量 time phasor
时间相位 time phase
时间谐波 time harmonic
时间延迟 TD (time delay)
时间整定 time setting
时间轴 time axis
时间坐标 time base
时控脉冲 timed pulse
时偏 time bias
时限低电压保护 time undervoltage protection
时限断路器 time-limit breaker
时限附件 time-limit attachment
时限过电流 time overcurrent
时限继电器 time-lag relay; time-limit relay
时限元件继电器 time-element relay
时延 TD (time delay)
时延伺服系统 time-delay servo (system)
时延随动系统 time-delay servo (system)
时域电磁学 time-domain electromagnetics
时域矩阵 time-domain matrix
时滞效应 time-lag effect
识别 identification
识别信标 identification beacon
实心导体 solid conductor
实心导线 solid conductor
实验电路板 breadboard

实用控制程序 UCP (utility control program)
拾音器 pickup
拾振器 vibration pickup
使用寿命〔安装后的〕 installed life
使用条件 service condition
示波法 oscillography
示波管 oscilloscope tube
示波计 oscillometer
示波器 oscillograph; oscilloscope
示波器探头 oscilloprobe
示波术 oscillography
示波图 oscillogram
示数器 numeroscope
示振器 kaleidophone
示踪气体 tracer gas
示踪元素 tracer element
示踪原子 tracer atom
势能 potential energy
事故备用电源 EPS (emergency power supply)
事故运行方式 emergency operating mode
事故自诊断 self-diagnosis of fault
事件树 event tree
试点电站阶段 pilot-plant stage
试电笔 electroprobe
试运行 TR (test run)
视差 parallax
视场 visual field
视程 visual range
视见函数 visible function
视距测定仪 visual range meter
视距仪 stadia
视频分布放大器 VDA (video distribution amplifier)
视网膜电描记术 electroretinography
视网膜电图 ERG (electroretinogram)
视网膜电图描记器 electroretinograph
视网膜电图学 electroretinography
视在功率表 volt-ampere meter
视在功率电度表 volt-ampere-hour meter
视轴 visual axis
是非决策 yes-no decision
适配器 adaptor
适时值 just value
适用性 serviceability
室内布置 indoor arrangement
室内接地开关 indoor earthing switch
室内照明 interior lighting
释能度 exoergicity
释能反应 exoergic reaction
收发器 transceiver
收发器数据链路 transceiver data link
收集器 catcher
收缩 shrinkage
收缩率 shrinkage
手动复位 hand reset
手控干预 intervention by manual control
受电弓 pantograph
受激络合物 excited complex
受激原子 excited atom
受油器 oil feeder
受主电离能 ionizing energy of acceptor
枢轴支架 pivot support frame
疏水阀 drain valve
输出导抗 output immittance
输出电路 output circuit
输出端数 fan-out
输出功率 watt-out
输出设备 output device
输出瓦数 wattage output
输出装置 output device
输电变压器 transmitting transformer
输电电压 transmission voltage
输电端 transmission end
输电范围 transmission range
输电路线 transmission route
输电能力 transmission capacity

输电塔 transmission tower
输电网 (transmission) grid; transmission network
输电系统 transmission system
输电线 transmission line; transmission wire
输电线防震锤 torsional damper
输电与配电 T & D (transmission and distribution)
输配电 T & D (transmission and distribution)
输入 input
输入板 tablet
输入变量 input variable
输入步长 input step
输入触发电压 input triggering voltage
输入单元 input unit
输入导抗 input immittance
输入电缆 incoming cable
输入电路 input circuit
输入电容 input capacitance
输入电压范围 input voltage range
输入电阻 input resistance
输入端 lead-in
输入分辨率 input resolution
输入功率 watt-in
输入激励量 input energizing quantity
输入检验 incoming inspection
输入开关 incoming circuit-breaker
输入馈线 incoming feeder
输入能量 imported energy
输入偏移电流 input offset current
输入绕组 input winding
输入设备 incoming unit
输入失调电流 input offset current
输入输出 input/output
输入输出电平 input/output level
输入输出对 input/output pair
输入速率 input transfer rate
输入提示 input prompt
输入调节系数 input regulation coefficient
输入阈电压 input threshold voltage
输入轴 input shaft
输入装置 input
输水管道 conduit
输送渠道 transfer canal
输送特性 transmission characteristic
输油泵 oil transfer pump
输运泵 transfer pump
输运平均自由程 transport mean free path
输运算符 transport operator
熟化 slaking
熟石灰 slaked lime
鼠笼式转子 squirrel-cage rotor
束带 bridle
束缚电 latent electricity
树脂 resin
树脂黏合石墨电刷 resin-bonded graphite brush
竖井 shaft
竖轴 vertical axis
数据电路端接设备 DCTE (data circuit terminating equipment)
数据电路终接设备 DCE (data circuit terminating equipment)
数据终端设备 DTE (data terminal equipment)
数量 magnitude; quantity
数模转换 D/A conversion (digital-to-analogue conversion)
数模转换器 DAC (digital-to-analogue converter)
数字电液控制系统 digital electro-hydraulic system
数字记录器 numeroscope
数字脉冲宽度调制 DPDM (digital pulse duration modulation)
数字配线架 DDF (digital distribution frame)
数字输入 DI (digital input)
衰变 decay

衰减 damping; decay; fade-out
衰减器 attenuator
衰减时间 fall time
闩锁继电器 lock-up relay
栓 key
栓钉 stud
双T形网络 twin-T network
双Δ twin-delta
双波绕组 duplex wave winding
双层绕组 two-layer winding
双冲程发动机 two-cycle engine
双触点 twin contact
双床除盐装置 two-bed demineralizer
双床的 twin-bed
双磁芯开关 two-core switch
双电极焊条 twin electrode
双电枢电机 two-armature motor
双电枢共轴电动机 tandem motor
双叠绕组 duplex lap winding
双二极管 twin diode
双工 duplex
双股线 twin wire
双回程锅炉 two-pass boiler
双回路伺服机构 two-loop servomechanism
双击式水轮机 tangential flow turbine
双极开关 two-pole switch
双绞股线 twin twisted strand
双绞线 twisted pair
双金属 bimetal
双金属腐蚀 bimetallic corrosion
双晶体三极管 tandem transistor
双颈蒸馏瓶 two-neck(ed) distilling flask
双壳冷凝器 twin-shell condenser
双壳凝汽器 twin-shell condenser
双力矩 bimoment
双联发动机 twin engine
双联绝缘子串 twin insulator strings
双联可变电容器 two-gang variable capacitor
双列汽缸发动机 two-bank engine
双流程堆芯 two-flow core; two-pass core
双流程凝汽器 two-pass condenser
双流汽轮机 twin-turbine
双流体反应堆 two-fluid reactor
双流体雾化 twin-fluid atomization
双流体系统 two-fluid system
双流型汽轮机 twin-turbine
双炉膛 twin furnace
双炉体锅炉 two-furnace boiler
双路冷却 two-pass cooling
双路绕组 two-circuit winding
双螺杆挤出机 twin-screw extruder
双面凹的 biconcave
双面凸的 biconvex
双膜理论 two-film theory
双汽轮发电机 twin-turbine generator
双枪示波器 two-gun oscillograph
双曲轴发动机 twin crankshaft engine
双群扩散理论 two-group diffusion theory
双群理论 two-group theory
双群临界方程 two-group critical equation
双群临界质量 two-group critical mass
双群模型 two-group model
双群微扰理论 two-group perturbation theory
双绕组变压器 two-winding transformer
双三极管 twin triode
双输入加法器 two-input adder
双输入减法器 two-input subtracter
双输入开关 two-input switch
双输入门 two-input gate
双输入伺服 two-input servo
双调谐电路接收机 two-circuit receiver

双铁芯励磁绕组　two-core excitation winding
双头螺栓　stud
双位置动作　two-position action
双线圈继电器　two-coil relay
双线圈仪表　two-coil instrument
双向拉伸　biaxial stretching
双向(三极)晶闸管　triac
双相的　bi-phase
双象限变流器　two-quadrant converter
双芯变压器　two-core transformer
双芯电缆　twin(-core) cable
双芯电流互感器　twin-core current transformer
双芯反应堆　two-core reactor
双星形接法　Y-Y connection
双针探针　twin probe
双支热电阻　duplex RTD (duplex resistance temperature detector)
双周波发电机　two-cycle generator
双轴承型电机　two-bearing machine
双柱式变压器　two-legged transformer
双柱型铁芯　two-column core
双组分热管　two-component heat pipe
水泵　water aspirator; water pump
水泵房　water pump house
水表　water meter
水表面玻璃　water-gauge glass
水波模拟　water-wave analogy
水槽蒸发器　water-bath evaporator
水侧腐蚀　waterside corrosion
水澄清作用　water clarification
水处理　water conditioning; water processing; water treatment
水处理厂　water treatment plant
水处理厂房　water treatment building
水处理化学制品　water treatment chemical
水处理设备　water treatment equipment
水处理设施　water treatment facility
水处理室　water treatment room
水处理系统　water treatment system
水锤　water hammer
水萃取介质　water-extracting medium
水当量　water equivalent
水道　water channel
水电厂　hydropower plant
水电池　water battery
水电站　hydropower station
水电阻器　water resistor
水动机　water motor
水反射系统　water-reflected system
水放射性测量计　water radioactivity meter
水分析　water analysis
水分盈余　water surplus
水封　water seal (gland)
水封室　water seal chamber
水封套　water-sealed gland
水封箱　water seal tank
水辐解　water radiolysis
水腐蚀　water corrosion
水负载　water load
水负载功率计　water-load power meter
水垢分析　scale analysis
水管　water pipe; water tube
水管锅炉　water tube boiler
水管接地制　water pipe grounding system
水管冷却　pipe cooling
水管式锅炉　water tube boiler
水锅炉中子源　WBNS (water boiler neutron source)
水化学　hydrochemistry
水环泵　water ring pump
水灰比　w/c (water-cement ratio)

水回收 water recovery
水回收设备 water recovery apparatus
水回用 water reuse
水活度测量仪 water activity meter
水击 water hammer
水加氯处理 water chlorination
水加热 water heating
水加热槽 water heating bath
水夹套 water leg
水假体 water phantom
水监测 water monitoring
水监测网 water surveillance network
水浸入 water ingress
水浸式刮板捞渣机 SSC (submerged scraper conveyor)
水浸试验 water immersion test
水净化 water purification
水控制阀 water control valve
水控装置 water control device
水库贮水量 pondage
水冷壁 water(-cooled) wall
水冷壁管 water wall tube
水冷壁回路 water wall circuit
水冷壁冷却 water wall cooling
水冷壁炉膛 water-walled furnace
水冷壁面 water wall surface
水冷变压器 water-cooled transformer
水冷淬火 water hardening
水冷电磁铁 water-cooled electromagnet
水冷电动机 water-cooled motor
水冷电机 water-cooled machine
水冷电抗器 water-cooled reactor
水冷电阻器 water-cooled resistor
水冷发电机 water-cooled generator
水冷反应堆 water-cooled reactor
水冷炉排 water-cooled grate
水冷炉膛 water-cooled furnace
水冷汽轮发电机 water-cooled turbogenerator
水冷汽轮机 water-cooled turbine
水冷强制油循环式变压器 water-cooled forced-oil transformer
水冷却器 water cooler
水冷却塔 water-cooling tower
水冷栅格 water-cooled lattice
水冷式冷凝器 water-cooled condenser
水冷系统 water-cooling system
水冷油浸变压器 water-cooled oil-immersed transformer
水冷油绝缘变压器 water-cooled oil-insulated transformer
水冷转子 water-cooled rotor
水力冲洗 water washing
水力除灰 water sluicing
水力除灰系统 water-sluicing system
水力发电站 hydropower station
水力发动机 water engine; water motor
水力喷射泵 water-jet pump
水力涡轮机 hydraulic turbine
水力引射器 water-jet fan
水力制动器 water brake
水流 water flow
水流继电器 water-flow relay
水流量警报系统 water-flow alarm system
水流量调节器 water-flow regulator
水轮发电机 hydrogenerator; water turbine generator; water turbogenerator; water-wheel generator
水轮发电机组 hydroelectric set
水轮机 hydraulic turbine
水慢化反应堆 water-moderated reactor
水慢化剂 water moderator
水慢化水冷反应堆 water-moderated water-cooled reactor

水煤气 water gas
水煤气转化 water gas shift
水密试验 watertight test
水密罩 watertight closure
水膜 water film
水膜除尘器 water-film scrubber
水膜系数 water film coefficient
水幕 water curtain
水喷射器 water ejector
水喷雾式灭火器 water spray fire extinguisher
水喷嘴 water-jet nozzle
水平分辨率 horizontal resolution
水平衡 water budget
水平奇偶校验 horizontal parity check
水平楼面接线电缆 horizontal floor wiring cable
水平清晰度 horizontal resolution
水平扫描频率 horizontal frequency
水平仪 leveller
水气比 water-air ratio
水汽 water vapour
水汽比 water-to-steam ratio
水汽含量 water vapour content
水韧处理 water toughening
水容量 water volume
水溶性 water solubility
水溶性缓蚀剂 water-soluble inhibitor
水溶性卤素 water-soluble halogen
水溶性树脂 water-soluble resin
水溶液 aqueous solution
水软化 water softening
水软化剂 water softening agent
水润滑轴承 water-lubricated bearing
水射流切割 water-jet cutting
水射器 water injector
水蚀 water erosion
水-水反应堆 water-water reactor
水-水换热器 water-to-water heat exchanger; water-water heat exchanger
水-水式换热器 water-water type heat exchanger
水锁 water blocking
水套 water jacket
水套冷凝器 water-jacketed condenser
水体污染 water body pollution
水通量 water flux
水头 water head
水位 water level
水位报警 water-level alarm
水位反应性系数 water-level reactivity coefficient
水位记录器 water-level recorder
水位监测仪表 water-level instrumentation
水位指示浮子 water-level float
水污染 water contamination; water pollution
水污染监测仪 water pollution monitor
水污染指数 water pollution index
水务管理 water management
水雾化器 water atomizer
水洗 water washing
水洗涤器 water scrubber
水系统 water system
水线腐蚀 waterline corrosion
水循环 water circulation
水循环泵 water circulation pump
水循环管 water circulation pipe
水循环回路 water-flow circuit
水循环检测器 water circulation detector
水循环系数 water circulation coefficient
水循环系统 water circulation system; water-cycling system
水压 water pressure
水压继电器 water pressure relay
水养护 water curing

水银继电器　mercury relay
水硬度　water hardness
水铀体积比　water-to-uranium volume ratio; water/uranium volume ratio
水铀原子比　water-to-uranium atomic ratio
水铀质量比　water-to-uranium mass ratio
水预处理　water pretreatment
水运　water carriage
水运系统　water-carriage system
水再循环　water recirculation
水载废物　water-borne waste
水闸　penstock; sluice
水蒸气　steam; water vapour
水蒸气电弧焊　water vapour arc welding
水蒸气渗透性　water vapour permeability
水蒸气吸收　water vapour absorption
水质标准　water quality standard
水质分析　water quality analysis
水质管理　water quality management
水质监测　water (quality) monitoring
水质控制　water quality control
水质目标　water quality goal
水质判据　water quality criterion
水质评价　water quality evaluation
水质污染　water (quality) pollution
水质污染物　water quality pollutant
水中放电成形　electrohydraulic forming
水柱　WC (water column)
水柱避雷器　water-column arrester; water-jet arrester
水柱校正　water-column correction
水柱压力　water-column pressure
水阻负载箱　water-resistance load tank
水阻器　water resistor
顺控　sequential control
顺控系统　sequential control system
顺序表图　sequence chart
顺序程序　sequential program
顺序分解　sequential decomposition
顺序控制　sequential control
顺序控制器　sequential controller
顺序控制系统　sequential control system
顺序控制站　sequential control station
顺序式自动光谱仪　sequential auto-spectrometer
顺序拓扑　sequential topology
顺序优化　sequential optimization
瞬变特性　transient behaviour
瞬变现象　transient phenomenon
瞬间磁链　transient flux linkage
瞬间峰值　transient peak
瞬接触点　snap-on contact
瞬时保护　instantaneous protection
瞬时电流　instantaneous current
瞬时动作　instantaneous operation
瞬时功率　instantaneous power
瞬时故障　transient fault
瞬时过载电流　transient overload current
瞬时可用度　transient availability
瞬时可用率　transient availability
瞬时强迫停运　transient forced outage
瞬时切断　instantaneous trip
瞬时热响应　transient thermal response
瞬时特性　transient property
瞬时脱扣器　instantaneous release
瞬时态　transient behaviour
瞬时阻力　transient drag
瞬衰电流　transient-decay current

瞬态 transient state
瞬态电抗 transient reactance
瞬态电流 transient current
瞬态电流补偿 transient current offset
瞬态电压 transient voltage
瞬态短路时间常数 transient short-circuit time constant
瞬态反应 transient response
瞬态沸腾 transient boiling
瞬态分量 transient component
瞬态分压器 transient divider
瞬态工况 transient condition
瞬态过电压 transient overvoltage
瞬态过电压计数器 transient overvoltage counter
瞬态恢复电压 transient recovery voltage
瞬态临界流 transient critical flow
瞬态脉冲 transient pulse
瞬态平衡 transient equilibrium
瞬态曲线 transient curve
瞬态热应力 transient thermal stress
瞬态热阻抗 transient thermal impedance
瞬态稳定(度) transient stability
瞬态响应 transient response
瞬态应力 transient stress
瞬态振荡 transient oscillation
瞬态振动 transient vibration
司太立合金 stellite (alloy)
丝极 filament
丝锥 tap
斯坦顿数 Stanton number
死区起始值 initial value of dead band
四重阀 quadruple valve
四极残余气体分析器 quadrupole residual gas analyzer
四极场 quadrupole field
四极磁铁 quadrupole magnet
四极杆 quadrupole rod
四极离子阱 quadrupole ion trap
四极滤质器 quadrupole mass filter
四极探头 quadrupole probe
四极质谱计 quadrupole mass spectrometer
四位字节 nibble
四位组 nibble
伺服电(动)机 relay; servomotor
伺服电动执行器 servomotor actuator
伺服放大器 servoamplifier
松脂 resin
送风风口 supply outlet
送风管道 supply duct
送风器 blower
送水 water delivery
搜集器 arrester
速动 quick action; quick operation
速动断路器 quick-operating circuit-breaker
速动继电器 quick-operating relay
速度 velocity
速度传感器 velocity pickup; velocity sensor; velocity transducer
速度反馈 velocity feedback
速度分布 velocity distribution
速度聚焦 velocity focusing
速度面积法 velocity-area method
速度式水表 velocity-type water meter
速度调制管 klystron
速度误差 velocity error
速度误差系数 velocity error coefficient
速度限制器 velocity limiter
速敏输出电压 speed-sensitive output voltage
速释继电器 quick-releasing relay
速调管 klystron; transit-time tube
速调管振荡器 klystron oscillator
酸量法 acidimetry

酸气 sour gas
随动式机械手 master-slave manipulator
随附信息组 trailer block
碎屑 clastic
隧道二极管 TD (tunnel diode)
隧道式炉 tunnel furnace
损耗(量) ullage
榫槽接合 tongue and groove joint
梭 shuttle
羧酸 carboxylic acid
缩短率 reduction of length
缩放 zoom
缩脉 vena contracta
索 cord
索环 becket
索具 tackle
索引存取 indexed access
锁 lockup
锁闭电路 locking circuit
锁存器 latch
锁定 lock-in; locking; lockout
锁扣机构 latching device
锁扣接触器 latched contactor
锁气器 airlock
锁位 lock-on
锁住电路 latching circuit

Tt

他励 separate excitation
塔架 tower
塔式回收系统 tower reclaiming system
塔型锅炉 tower boiler
塔影效应 tower shadow effect
台面 mesa
台面晶体管 mesa transistor
台面刻蚀 mesa etching
太阳能 solar energy
太阳能光电转换 solar photovoltaic conversion
钛酸盐 titanate
泰勒标准筛 Taylor standard screen
泰勒公式 Taylor formula
泰勒积累因子 Taylor build-up factor
泰勒级数 Taylor series
泰勒接线法 Taylor connection
泰勒展开 Taylor expansion
弹簧片 leaf
弹簧压强 spring pressure
弹踢器 kicker
弹性触头 spring contact
弹性模量 elastic modulus
弹性模数 elastic modulus
弹性碰撞 elastic collision
弹性轴 elastic axis
弹黏性 elasticoviscosity
钽电容器 tantalum capacitor
钽整流器 tantalum rectifier
炭 charcoal
探测器 detector
探管传声管 probe microphone
探头摆动扫查 probe rotational scan
探头-缺陷距离 probe-to-flaw distance
探头线圈 probe coil
探头线圈间隙 probe coil clearance
探针 feeler
探针法 probe method
探针离子 probe ion
碳 carbon
碳棒 carbon rod
碳钢 carbon steel
碳化 carbonization
碳化物 carbide
碳素钢 carbon steel
碳酸定量器 kalimeter
碳酸氢盐 bicarbonate
碳酸盐 carbonate
碳锌电池 zinc-carbon battery
汤姆孙电桥 Thomson bridge
汤森放电 Townsend discharge
陶器 ceramics
陶土 kaolin
套管式电流互感器 BCT (bushing current transformer)
套筒 cartridge; sleeve
套装转子 shrink-on disc rotor
特低频 ULF (ultra-low frequency)
特定程序计算机 target computer
特定用途集成电路 ASIC (application specific integrated circuit)
特高频 UHF (ultra-high frequency)
特高频相关器 UHF correlator
特高压 UHV (ultra-high voltage)

特高压输电 UHV transmission
特高压输电线路 UHV transmission line
特勒根定理 Tellegen's theorem
特殊紧固件 special fastener
特殊余摆线 special trochoid
特细玻璃丝包线 ultrafine glass-coated wire
特细粉尘 ultrafine dust
特细粒子 ultrafine particle
特细煤烟 ultrafine soot
特细漆包线 ultrafine enamelled wire
特细微尘 ultramicroscopic dust
特种信号灯 optiphone
梯 ladder
梯塞绕组 teaser winding
梯塞线圈 teaser coil
梯式滤波器 ladder filter
梯形槽 trapezoid-shaped slot
梯形电路 ladder circuit
梯形桁架 trapezoid truss
梯形畸变 keystone distortion
梯形失真 keystone distortion
梯形图 ladder diagram
梯形网络 ladder network
梯形效应 ladder effect
提取 abstraction
提升滑车 hoisting tackle
提升机构 hoisting gear
提升间 hoistway
体电导式湿度传感器 volume conductive humidity sensor; volume conductive humidity transducer
体积流量 volume flow rate
体积流量总量 volume flow
体积黏度 volume viscosity
体积色谱法 volumetric chromatography
体膨胀法 volume thermodilatometry
体系 system
体系结构 system architecture
天然焦 carbonite
天线杆 mast

天线扫掠 lobing
天线射束控制 lobing
添加剂 additive
添加物 additive
甜气 sweet gas
填充方式 fill mode
填料 stuffing
填料函型电磁阀 packed-type solenoid valve
填料函组件 packing box assembly
填隙片 shim
挑坎 trajectory bucket
调节 adjustment
调节棒 adjuster rod
调节动作定律 law of regulating action
调节剂 conditioner
调节器 conditioner
调零 zero adjustment
调速发电机 velodyne
调速器 speed governor; speed regulator
调压变压器 joystick transformer
调压井 surge tank
调压塔 surge tower
调整 adjustment
调整线圈 trim coil
调制 modulation
调制传递函数 MTF (modulation transfer function)
调制器 modulator
跳变频率 jump frequency
跳动 jitter
跳线 jumper
跳线弛度 jumper sag
跳线分布 jumper assignment
跳线线夹 jumper clamp
跳转操作 jump operation
铁板 iron sheet
铁磁电动式比值计 iron-cored ferrodynamic ratio meter
铁磁录波器 ferromagnetic oscillograph
铁电性 ferroelectricity
铁轭 yoke

铁耗 iron loss
铁芯 core
铁芯电动仪表 iron-cored electrodynamic instrument
铁芯电抗器 iron-core reactor
铁芯松弛 slackening of iron core
铁芯线圈 iron-core coil
铁氧体 ferrite
停电时间 interruption duration
停堆 fast scram
停机 outage
停机时间 idle time
停闪频率 fusion frequency
挺杆传动装置 tappet gear
通电 energizing
通断 switch
通断的 make-and-break
通断作用 on-off action
通风干湿表 ventilated psychrometer
通风井 shaft
通风率 ventilation rate
通风帽 cowl
通风器 ventilator
通风温度表 ventilated thermometer
通量 flux
通路 path
通气管 airway
通气孔 air-breather
通气装置 air-breather
通态 on-state
通态电流 on-state current
通态电压 on-state voltage
通态耗散功率 on-state power dissipation
通信电缆 telecommunication cable
通信量 traffic
通信业务分配器 traffic distributor
通用分类系统 UCS (universal classification system)
通用分流电路 general-purpose branch circuit
通用分路 general-purpose branch circuit
通用符号 general symbol
通用缓冲控制器 UBC (universal buffer controller)
通用换算器 multiscaler
通用接口总线 GPIB (general-purpose interface bus)
通用开关 general-purpose switch
通用逻辑块 ULB (universal logic block)
通用偶极子 UD (universal dipole)
通用熔断器 general-purpose fuse
通用十进制分类法 UDC (universal decimal classification)
通用钨丝灯泡 general-service tungsten filament lamp
通用现场通信系统 general-purpose field communication system
通用异步接收发送器 UART (universal asynchronous receiver/transmitter)
通用异步接收发送设备 UART (universal asynchronous receiver/transmitter)
通用自动控制及测试设备 UACTE (universal automatic control and test equipment)
通用字符集 UCS (universal character set)
同步 synchronism
同步磁道 timing track
同步机 synchronizer
同步控制 synchronous control
同步器 synchronizer
同步示波器 oscillosynchroscope
同步相 locking phase
同步指示器 synchroscope
同步装置 synchronizer
同时比较法 simultaneous comparison method
同时并用技术 simultaneous technique

同时联用技术 simultaneous technique
同相电流 in-phase current
同相放大器 non-inverting amplifier
同相控制 in-phase control
同相连接 non-inverting connection
同相零位电压 in-phase null voltage
同相升压机 in-phase booster
同相输入 non-inverting input
同心式线圈 concentric coil
同轴电缆 coaxial cable
同轴开关 gang switch
铜 copper
铜铝合金 aluminium bronze
铜皮线 tinsel conductor
桶内贮存 tank retention
筒 cylinder
筒仓 silo
投光灯 floodlight
投光照明 floodlight
投运试验 commissioning test
透明 transparency
透明度 limpidity; transparency
透射比 transmission factor; transmittance
透射率 transmissivity
透射系数 transmission coefficient
透射因数 transmission factor; transmittance
透水(性) water penetration
凸 convex
凸极 protruding pole; salient pole
凸角 lobe
凸嵌线 bolection
凸缘 lug
突然失灵 glitch
突震 kick

图 drawing
图钉 tack
图像传感器 image sensor
图像寄存器 image register
图像增强 image enhancement
涂膏式极板 pasted plate
涂蜡纱包线 waxed cotton-covered wire
涂料 facing
土坯砖 adobe
土壤侵蚀 soil erosion
土壤下降 subsidence
湍流 turbulence; turbulent flow
团块 agglomerate
团粒 agglomerate
推斥感应电动机 repulsion induction motor
推力盘 thrust collar
推力套筒轴承 Jordan bearing
推力轴承 thrust bearing
推送运行 propelling movement
退火 annealing
退激 de-excitation
退水渠 waste canal
托架 bearer
托里拆利真空 Torricellian vacuum
托木 bolster
托普勒电机 Toepler machine
拖尾峰 tailing peak
拖曳电缆 trailing cable
脱甲烷塔 demethanizer
脱硫 desulphurization
脱溶硬化合金 precipitation hardened alloy
脱水 dehydration; water removal
脱水剂 dehydrant
脱水作用 dehydration
脱戊烷塔 depentanizer
脱盐 demineralization
脱氧 deoxygenation
拓扑分析 topological analysis

Ww

瓦 watt
瓦垄板 corrugated steel sheet
瓦秒 watt-second
瓦时 watt-hour
瓦时常数 watt-hour constant
瓦时计 watt-hour meter
瓦时容量 watt-hour capacity
瓦时效率 watt-hour efficiency
瓦时需量计 watt-hour demand meter
瓦数 wattage
瓦数损耗 wattage dissipation
瓦特计 wattmeter
瓦特计式继电器 wattmetric relay
瓦特小时 watt-hour
外摆线 epicycloid
外部磁感应影响 influence of magnetic induction of external origin
外部绝缘 EI (external insulation)
外对流敏感元件传感器 sensor with external convection sensitive element
外观检验 visual inspection
外过电压 external overvoltage
外护套 skin casing
外加 impress
外加电流 impressed current
外加电流保护 impressed-current protection
外加电流阳极 impressed-current anode
外加电压 impressed voltage
外加剂 admixture
外绝缘 EI (external insulation)
外壳鼓胀 swelling of case
外围 periphery
外围接口适配器 PIA (peripheral interface adaptor)
外围设备 peripheral device; peripheral equipment
外延层 epilayer
外延膜 epifilm
外引式接地 leading-out grounding
弯管 elbow
弯管机 tube bender
弯管接头 elbow
弯扭叶片 twisted blade
弯曲 curvature
弯曲度 camber
弯曲强度 bending strength
弯头 elbow; knee bend
完全二次型方根组合法 CQC (complete quadratic combination)
完全气体 perfect gas
完全真空 perfect vacuum
顽磁 magnetic remanence
万位 myriabit
万向接头 Hooke's coupling; knuckle joint
万向联轴器 Hooke's coupling
万用表 multimeter; multitester; universal instrument
网格 grid
网格点 grid point
网格接地体 grid-type earth electrode
网格栅 grid
网格式接地极 grid-type earth electrode
网格坐标 grid coordinate

网格坐标系 grid coordinate system
网关 gateway
网际互连协议 internetworking protocol
网孔 mesh
网孔电流 mesh current
网络布置 network layout
网络故障 network fault
网络控制 network control
网络模拟 network analogue
网状发射极 mesh emitter
往复式压缩机 reciprocating compressor
危险电位 hazard potential
微安 microammeter
微安表 microammeter
微安培 microampere
微巴 microbar
微波 microwave
微波发射器 microwave emitter
微波激射(器) maser
微波集成电路 MIC (microwave integrated circuit)
微波纵联保护 microwave-pilot protection
微尘学 koniology
微电路 MC (microcircuit)
微动 inching; jogging
微动控制 inching control; jog control
微动速度 jogging speed
微分 differential
微分控制器 D-controller; derivative controller
微分器 differentiator
微分作用 D-action; derivative action
微伏表 microvoltmeter
微伏(特) microvolt
微合金晶体管 MAT (microalloy transistor)
微矩阵 micromatrix
微孔结构 cellular structure
微控 inching control
微库(仑) microcoulomb

微粒 particulate
微量元素 microelement
微脉冲发生器 micropulser
微模块 micromodule
微姆(欧) micromho
微欧(姆) microohm
微瓦(特) microwatt
微隙开关 micro-gap switch
微型电位计 micropotentiometer
微型晶片 microwafer
微型开关 microswitch
微型元件 microelement
微型组件 microcomponent
微压计 micromanometer
韦伯 Wb (weber)
围板 baffle
围带 shroud
维持时间 hold time
维持因数 MF (maintenance factor)
维弧 keep-alive
维护 maintenance
维护系数 MF (maintenance factor)
维氏硬度计 Vickers hardness tester
维氏硬度压头 Vickers hardness penetrator
维氏硬度值 Vickers hardness number
维修 maintenance
维修度 maintainability
维修性 maintainability
尾部受热面 tail heating surface
尾冲波 trailing shock wave
尾端 tail end
尾端效应 tail effect
尾激波 trailing shock wave
尾矿 tailing(s)
尾料流 tail stream
尾水廊道 tailwater gallery
尾水渠 tail channel; tailrace
尾水隧洞 tailrace tunnel
尾水调压室 tailrace surge chamber
尾水闸门 tail gate

尾随涡　trailing vortex
尾缘激波　trailing-edge shock
尾渣　tailing(s)
纬度　latitude
纬度效应　latitude effect
未遂事故　near accident
位能　potential energy
位误差率　BER (bit error ratio)
位移　displacement
位移距离显示　K-display
位置编码器　position encoder
位置测量仪　position measuring instrument
位置传感器　position sensor; position transducer
位置反馈　position feedback
位置简图　location diagram
位置误差　position error
位置误差系数　position error coefficient
位置指示开关　position indicating switch
温差　TD (temperature difference)
温差电偶　TC (thermal couple; thermocouple)
温度补偿　TC (temperature compensation)
温度补偿电机　temperature-compensated electric machine
温度垂直廓线辐射仪　VTPR (vertical temperature profile radiometer)
温度-电阻曲线　temperature-resistance curve
温度控制器　TC (temperature controller)
温度控制装置　TCU (temperature control unit)
温度系数　temperature coefficient
温排水　warm water discharge
文丘里管　Venturi tube
文丘里喷嘴　Venturi nozzle
纹波　ripple
纹波电压　ripple voltage

紊流　turbulence; turbulent flow
稳定(化)　stabilization
稳定剂　stabilizer
稳定器　stabilizer
稳定着火　stable ignition
稳定作用　stabilization
稳燃器　firing stabilizer
稳态　steady state
稳压电路　voltage stabilizing circuit
涡　eddy
涡带　scroll
涡动速度　eddy velocity
涡街　vortex street
涡街流量传感器　vortex-shedding flow transducer
涡街流量计　vortex-shedding flowmeter
涡流　eddy current
涡流电动机　eddy current motor
涡流电路　eddy current circuit
涡流扩散　eddy diffusion
涡流损耗　eddy current loss
涡流制动　eddy current braking
涡流阻尼　eddy current damping
涡轮出口温度　TOT (turbine outlet temperature)
涡轮(机)　turbine
涡轮膨胀机　turbine expander
涡轮汽缸　turbine cylinder
涡轮压缩机　turbo-compressor
沃德-伦纳德系统　Ward-Leonard system
沃克普式起动器　Wauchope type starter
卧式电机　horizontal machine
乌尔夫静电计　Wulf electrometer
污染　contamination
污染物　contaminant; pollutant
污水　sewage; sewerage
污水管　waste pipe
污水坑　sump
钨准直器　tungsten collimator
无变压器电源　transformerless power supply

无变压器多接开关 transformerless multiplex switch
无衬砌隧洞 unlined tunnel
无触点继电器 contactless relay; non-contact relay
无电极放电 electrodeless discharge
无电抗电阻 non-reactive resistance
无电抗功率 non-reactive power
无电压 no-voltage
无电压继电器 no-voltage relay
无电压释放 no-voltage release
无感电路 non-inductive circuit
无感电容器 non-inductive capacitor
无感电阻 non-inductive resistance
无感分流器 non-inductive shunt
无功电度表 var-hour meter; watt-hour meter; wattless component
无功电流 wattless current
无功分量表 wattless component meter
无功功率 magner; reactive power
无功功率表 varmeter; wattless power meter
无故障工作时间 time between failures
无规律转换 irregular transition
无火花运行 sparkless operation
无级变速器 infinitely variable speed transmission
无级变速箱 infinitely variable speed transmission
无键插座 keyless socket
无接触的 non-contact
无接点的 non-contact
无菌的 aseptic
无量纲参数 non-dimensional parameter
无量纲系数 non-dimensional coefficient
无黏性煤 yolk coal
无损分解 non-loss decomposition
无通信监控 inactivity control
无湍流 non-eddying flow
无涡流 non-eddying flow
无限冲激响应 IIR (infinite impulse response)
无限大母线 infinite bus
无限脉冲响应 IIR (infinite impulse response)
无限增益 infinite gain
无线电测向仪 radio direction finder
无线电电子管 radio tube
无线电经纬仪 radio theodolite
无线电探空仪 radiosonde
无效接收 invalid reception
无旋场 irrotational field
无烟煤 anthracite
无因次参数 non-dimensional parameter
无因次系数 non-dimensional coefficient
无引线外壳 leadless package
无引线芯片载体 LLCC (leadless chip carrier)
无源电路 passive circuit
五二码 quibinary code
戊烷 pentane
戊烷馏除器 depentanizer
物镜 objective (lens)
物距 object distance
误比特率 BER (bit error ratio)
误触发 false triggering
误动作 misoperation
误跳闸 mis-trip
误通 false firing
误用失效 misuse failure
雾灯 fog lamp

吸附 adsorption
吸附剂 absorbent
吸流变压器 booster transformer
吸盘 magnetic chuck
吸入高度 suction height
吸入压力 inlet pressure
吸入压头 suction head
吸上高度 suction height
吸收材料 absorber
吸收剂 absorbent
吸收制冷 absorption refrigeration
吸水率 water absorptivity
吸水性 water absorption
析因实验 factorial experiment
稀释剂 diluent
稀土元素 rare earth element
熄灭 quenching
熄灭电路 quench circuit
熄灭电压 quenching voltage
洗涤塔 scrubber
系数 coefficient
系统 system
系统辨识 system identification
系统不确定度 systematic uncertainty
系统参数 system parameter
系统测试时间 system test time
系统方法 system approach
系统仿真 system simulation
系统分解 system decomposition
系统分析 system analysis
系统工程 systems engineering
系统管理 system management
系统规划 system planning
系统环境 system environment

系统集结 system aggregation
系统间干扰 intersystem interference
系统间故障 intersystem fault
系统建模 system modelling
系统矩阵 system matrix
系统可靠性 system reliability
系统可维护性 system maintainability
系统理论 system theory
系统灵敏度 system sensitivity
系统模拟 system simulation
系统模型 system model
系统模型化 system modelling
系统内部干扰 intra-system interference
系统偏差 system deviation
系统评价 system assessment; system evaluation
系统软件 system software
系统设计说明 SDD (system design description)
系统生产时间 system production time
系统同构 system isomorphism
系统同态 system homomorphism
系统统计分析 system statistical analysis
系统误差 systematic error
系统性能试验 system performance test
系统学 systematology
系统优化 system optimization
系统诊断 system diagnosis
系统中断 system interrupt
系统状态 system state
系统资源 system resource

系统阻抗比 system impedance ratio
系统最高电压 highest voltage of a system
系统最优化 system optimization
细调 fine adjustment
瑕疵 blemish
下标 subscript
下风式风电机 downwind WTG
下降管 downcomer
下降时间 fall time
下漏 underflow
下水道 sewer
先导 leader
氙棒组 Xe-rod bank
氙和钐中毒 Xe and Sm poisoning
氙积累 xenon accumulation; xenon build-up
氙控制组位置控制回路 Xe-bank position control loop
氙振荡 xenon oscillation
氙中毒 xenon poisoning
衔铁 armature
显示板 indicator board
显像管 kinescope; oscillight
现场浇注混凝土 in-situ concrete
现场平衡 in-situ balancing
现场维修 in-situ maintenance
现场用智能设备 intelligent field device
限定设备 qualified facilities
限额 quota
限幅器 limiter
限幅器电路 limiter circuit
限弧件 muffler
限流电路 limited current circuit
限流电阻 limiting resistance
限流阀 restrictor
限流器 current limiter
限位器 stop
限温器 temperature limiter
限值 limiting value
限制负载电阻(器) LLR (load limiting resistor)
限制器 killer; limiter

线棒 (winding) bar
线槽 slot
线电压监控器 LVM (line voltage monitor)
线路端子 line terminal
线路发报机 LT (line transmitter)
线路激励放大器 LDA (line driving amplifier)
线路集中器 LC (line concentrator)
线路末端升压器 tail-end booster
线路图 hook-up
线圈 coil
线圈间绝缘 intercoil insulation
线圈〖调节变压比的〗 teaser coil
线式调制 LTM (line-type modulation)
线外编码 out-of-line coding
线芯 core
线性 linearity
线性电路 linear circuit
线性电阻 linear resistance
线性度 linearity
线性化电路 linearizing circuit
线性化电阻 linearizing resistance
线性集成电路 linear integrated circuit
线性技术 linear technology
线性可变变压器 LVT (linear variable transformer)
线性控制 linearity control
线性能量转移 LET (linear energy transfer)
线性膨胀 linear expansion
线性闪光管 LFT (linear flash tube)
线性算符 linear operator
线性算子 linear operator
线性调制器 linear modulator
线性误差 linearity error
线性系统仿真 linear system simulation
线轴 drum stand; spool
线轴式绝缘体 spool insulator

陷波电路 trap circuit; wave trap
陷波功率计 notch power meter
陷波频率 notch frequency
相对测振法 vibration measurement by relative method
相对电容率 relative permittivity
相关图 correlation diagram
相关总线信号 interlocked bus signal
相互调制 intermodulation
相互作用 interaction
相切条件 tangency condition
相容性 compatibility
相似定律 law of similarity
箱底加热器 tank bottom heater
箱式断路器 tank-type circuit-breaker
箱式(反应)堆 tank-type reactor
箱液面指示器 tank level indicator
向电性 electrotropism
向斜 syncline
项目代号 item designation
巷道 lane
相 phase
相变点 transformation temperature
相变温度 transformation temperature
相差 phase difference
相电压 Y-voltage
相间变压器 interphase transformer
相间电抗器 interphase reactor
相间短路 interphase short circuit
相间绝缘 interphase insulation
相绕组 phase winding
相位 phase
相位标记 phase mark
相位差 phase difference
相(位)移 phase displacement
象限 quadrant
象限静电计 quadrant electrometer
象限仪 quadrant
肖氏硬度计 Scleroscope
消波装置 wave absorber
消除器 suppressor
消电离 deionization
消毒剂 disinfectant
消毒器 sterilizer
消火花电路 quenching circuit
消声器 muffler; noise eliminator; silencer
消隐脉冲 quenching pulse
硝基漆 lacquer
硝基漆层 lacquer layer
销 key
小波 wavelet
小步走刀 fine feed
小齿轮 pinion
小岛效应 island effect
小功率变压器 wattage transformer
小键盘 keypad
小孔 pore
小型电动机车 mule
小型开关 mini-switch
效率 efficiency
效应 effect
斜槽因数 skew factor
斜撑 knee bracing
斜道 shoot
斜垫轴承 tilting-pad bearing
斜度 obliquity
斜角 bevel
斜面 bevel
斜楔 taper wedge
谐波 harmonic
谐波补偿 harmonic compensation
谐波发生器 harmonic generator
谐波分量 harmonic component
谐波分析 harmonic analysis
谐波分析仪 harmonic analyzer; wave analyzer
谐波检波器 harmonic detector
谐波滤波器 harmonic filter
谐波试验 harmonic test

谐波稳定 harmonic restraint
谐波吸收器 harmonic absorber
谐波谐振 harmonic resonance
谐波抑制 harmonic restraint
谐波指示器 harmonic detector
谐和力波 harmonic force wave
谐振 resonance
谐振电路 tank circuit
写缓冲器 WB (write buffer)
泄地电流 EC (earth current); ground current
泄放阀 relief valve
泄放水净化 trap-water purification
泄放旋塞 waste cock
泄漏 leakage
泄漏电流 leakage current
泄漏放电 leakage discharge
泄漏压力 leak pressure
泄水锥 runner cone
芯柱 stem
锌 zinc
信标 beacon
信关 gateway
信号 signal
信号持续时间 signal duration
信号处理 signal processing
信号处理系统 signal processing system
信号电缆 signal cable
信号电路 signal circuit
信号电平 signal level
信号幅值顺序控制 signal amplitude sequencing control
信号复示器 signal repeater
信号隔离 signal isolation
信号检测和估计 signal detection and estimation
信号流(程)图 signal flow diagram
信号选择器 signal selector
信号转换器 signal converter
信噪比 signal-to-noise ratio
星绞 quadding
星绞机 star quadding machine
星形电压 wye voltage
星形接法相绕组 Y-connection phase winding
星形连接 star connection
星形连接电阻箱 Y box
星形联结 Y-connection; Y-junction
星形轮 spider
星形-三角形电力变压器 wye-delta power transformer
星形-三角形连接(法) wye-delta
星形-三角形起动器 Y-delta starter
星形四线组 star quad
行波 travelling wave
行波场 travelling field
行波管 TWT (travelling wave tube)
行波速调管 twystron
行波天线 wave antenna
行程 stroke; travel
行程开关 travel switch
行移场 travelling field
行移过电压 travelling overvoltage
形状因数 form factor
型线轧拉机 wire flattening and profiling machine
型效率 type efficiency
性能频带 performance band
袖珍变压器 ouncer transformer
溴化物 bromide
虚部 wattless component
虚拟存储(器) virtual memory
虚像 virtual image
虚像质谱计 virtual image mass spectrometer
序贯最小二乘估计 sequential least squares estimation
续流臂 free-wheeling arm
蓄电池 accumulator; EPS (electric power storage); secondary cell; storage cell
蓄热器 regenerator
蓄压器 accumulator
悬臂 boom
悬臂梁 cantilever

悬臂托梁 ancon
悬垂绝缘子 tie-down insulator
悬吊式锅炉构架 suspended boiler structure
悬浮 suspension
悬浮固体 SS (suspended solid)
悬浮微粒 aerosol
悬浮物 seston; SS (suspended solid)
悬浮液 suspension
悬链线 catenary
悬式绝缘子 suspension insulator
旋风分离器 cyclone
旋光性 opticity
旋光仪 polarimeter
旋进流量计 vortex precession flowmeter
旋流器 swirler
旋钮 knob
旋塞阀 cock
旋涡 eddy
旋涡流量计 vortex flowmeter
旋转 revolution; slewing
旋转备用 spinning reserve
旋转开关式电阻箱 resistance box with rotary switch
选件 option
选通电路 gate circuit
选通控制 gate control
选通脉冲 strobe pulse
选项 option
选择 option
选择开关 option switch
雪载 snow load
循环 circulation; loop
循环变频器 cycloconverter
循环流化床 CFB (circulating fluidized bed)
循环器 circulator
循环冗余校验 CRC (cyclic redundancy check)
循环色谱法 recycle chromatography
循环水泵 water circulating pump

压电 piezoelectricity
压控晶体振荡器 VCCO (voltage-controlled crystal oscillator); VCXO (voltage-controlled X-tal oscillator)
压力钢管 penstock
压力管道 penstock
压力控制阀 PCV (pressure control valve)
压力式涡轮 pressure turbine
压力损失 pressure loss
压力铸造 die casting
压敏变阻器 varistor
压敏电阻器 VDR (voltage-dependent resistor)
压片 sheeting
压片涂层 tablet coating
压气机 compressor
压铅 lead extrusion
压强计 manometer
压实 compaction
压水(反应)堆 PWR (pressurized water reactor)
压缩冲程 compression stroke
压缩点火 compression ignition
压缩机 compressor
压缩空气高速断路器 high-speed air-blast breaker
压铸 die casting
亚临界 subcritical
亚临界压力锅炉 subcritical pressure boiler
亚临界压力汽轮机 subcritical pressure turbine
亚稳能级 metastable level
亚稳态 metastable state
氩 argon

烟囱 stack
烟粒 soot
烟煤 bituminous coal
烟密度计 kapnometer
烟气 flue gas; stack gas
烟气排放控制 flue gas emission control
烟气脱氮 flue gas denitrification
烟气脱硫 flue gas desulphurization
延长器 extender
延迟定理 lag theorem
延迟值 length of delay
延度 ductility
延时电路 time-delay circuit
延时复位按钮 time-delay push button
延时继电接触器 time-delay contactor relay
延时继电器 time-delay relay; time-element relay; time-lag relay
延时间隔继电器 interval time-delay relay
延时欠电压继电器 time-delay undervoltage relay
延时熔断器 time-lag fuse
延时释放(器) time release
延时停机继电器 time-delay stopping relay
延性 ductility
岩溶 karst
岩石密度计 rock densitometer
岩石压入硬度计 rock press-in sclerometer
沿程损失 pipeline loss
研磨 lap

研磨剂 lapping compound
研磨模 lap
研磨硬度 lapping hardness
盐化作用 salinization
盐水 brine
衍射 diffraction
掩模 mask
掩模窗 mask hole
掩模架 mask holder
掩模组 mask set
眼钩 eye hook
验潮杆 tide pole
验潮仪 tide gauge; tide-meter
验磁器 magnetoscope
验电法 electroscopy
验电器 electroscope
阳电性 electropositivity
阳极检波 transrectification
阳极检波器 transrectifier
阳极检波系数 transrectification factor
阳离子 anode; cation
阳离子交换树脂 cation exchange resin
杨氏模量 Young's modulus
杨氏模数 Young's modulus
杨氏弹性模量 YME (Young's modulus of elasticity)
仰角 elevation
氧合作用 oxygenation
氧化 oxidation
氧化钙 lime
氧化还原电位测定仪 redox potential meter
氧化还原复合电极 redox electrode assembly
氧化剂 oxidant; oxidizer
氧化铝 aluminium oxide
氧化钛 titanium dioxide
氧化物 oxide
样条 spline
摇摆曲线 swing curve
摇表 tramegger
遥控 telecontrol
钥匙 key
冶金学 metallurgy
冶炼厂 smelter
业务量 traffic
叶顶间隙 tip clearance
叶尖 tip of blade
叶尖速比 tip speed ratio
叶轮 impeller
叶片 vane
叶片型电路 vane-type circuit
页面寻址 page addressing
页式打印机 page-at-a-time printer; page printer
页岩 shale
液滴夹带 droplet entrainment
液力涡轮机 hydraulic turbine
液体电阻器 liquid resistor
液体绝缘 liquid insulation
液体绝缘套管 liquid insulated bushing
液体阻尼器 liquid damper
一般符号 general symbol
一般公差 general tolerance
一次电池 primary cell
一次风 primary air
一次绕组 primary winding
一次一页式打印机 page-at-a-time printer
一致数组 UA (uniform array)
一致阵列 UA (uniform array)
一致自动数据处理系统 UADPS (uniform automatic data processing system)
伊尔格纳发电机组 Ilgner generator set
伊尔格纳系统 Ilgner system
仪表安全电流 instrument security current
仪表保安因数 instrument security factor
仪表常数 meter constant
仪表额定极限一次电流 rated instrument limit primary current
仪表盒 instrument case
仪表盘 fascia
仪表箱 instrument case
仪器 apparatus

移动起重机 travelling crane
移动式吹灰器 travelling soot blower
移动式卷扬机 travelling hoist
移动式模板 travelling shuttering
移动式启闭机 travelling hoist
移动相位 travelling phase
移动装置 shifter
移位 shifting
移位器 shifter
移液管 transfer pipet
异步电动机 non-synchronous motor
异步起动器 induction starter
异常事件记录系统 UERS (unusual event recording system)
异相电流 out-of-phase current
异形导线 shaped conductor
异性电 opposite electricity
抑制 suppression
抑制布线技术 suppressive wiring technique
抑制电路 killer circuit
抑制电容器 suppression capacitor
抑制剂 inhibitor; suppressant; suppressor
抑制元件 suppression component
译码器 decipherer
意外接触 inadvertent contact
意外事故 contingency
意外事件 contingency
溢洪道 spillway
溢流 overrun
翼墙 wing wall
阴电子 negatron
阴极 cathode; negative electrode
阴极射线管 CRT (cathode ray tube)
阴接 female contact
阴离子 anion; negative ion
阴像 negative image
音频发生器 tone generator
音频振荡器 tone generator
音速 sonic speed
音响测深自动记录仪 echograph
引出端 leading-out terminal
引导电弧 pilot arc
引发剂 initiator
引风机 induced draft fan
引弧 arc initiation
引火 priming
引燃电流 ignition current
引燃电路 ignition circuit
引燃管 ignitron
引入 lead-in
引入线 lead-in
引射器 ejector
引线 lead; leader; lead (wire); leg
引线电缆 leader cable
引线键合 lead bonding
引线图案 lead pattern
隐极 non-salient pole
印刷电路板 PCB (printed circuit board)
英国标准协会 BSI (British Standards Institution)
英(制)热单位 Btu (British thermal unit)
荧光分光光度计 spectrofluorophotometer
影响系数 influence coefficient
应变 strain
应变仪 strainmeter
应电性 electrotropism
应急照明 emergency lighting
应力腐蚀 stress corrosion
应力集中 stress concentration
应力计 stress gauge
应用参考数据 ARD (application reference data)
应用管理中心 UCC (utility control center)
硬度计 sclerometer
硬件故障 hardware failure
硬煤 anthracite
硬石 adamant
硬挺度 stiffness
硬橡胶 ebonite

拥塞　congestion
永磁动圈式检流计　permanent-magnet moving coil galvanometer
永磁动圈式仪表　permanent-magnet moving coil instrument
永磁发电机　magneto
永磁透镜　permanent magnetic lens
永久性故障　permanent fault
涌流　inrush (transient) current
涌流抑制　inrush restraint
用户输入输出设备　UIOD (user I/O device)
用户属性数据集　UADS (user attribute data set)
用户线　in-house line
用键固定　key
用量　dose
优化　optimization
优化程序　optimizer
优先定向　preferred orientation
优质燃料　premium fuel
油顶起轴承　oil-jacked bearing
油断路开关　OCB (oil-circuit-breaker)
油断路器　OCB (oil-circuit-breaker)
油灰　badigeon
油浸电容器　oil condenser
油绝缘　oil insulation
油开关　oil switch
油冷变压器　tank transformer
油泥　sludge
油箱冷却器　tank cooler
油箱式变压器　tank transformer
油压马达　oil hydraulic motor
油制动器　oil brake
油阻尼　oil damping
游标　nonius; vernier
有齿电枢　toothed armature
有电部件　live part
有功部分　watt component; wattful component
有功电流　active current; wattful current
有功分量　watt component; wattful component
有功功率　non-reactive power; wattful power
有机半导体　organic semiconductor
有机酯　organic ester
有溶解力的　solvent
有色玻璃　stained glass
有限冲激响应　FIR (finite impulse response)
有限转角力矩电机　limited angle torque motor
有效电感　effective inductance
有效电流　effective current; wattful current
有效电压　effective voltage
有效负载　effective load
有效截面　effective cross-section
有效热　useful heat
有效输出　effective output
有效性　soundness
有效值　effective value
有效阻抗　effective impedance
有眼螺母　eye nut
有眼螺栓　eye bolt
有源电路　active circuit
有载电压　on-load voltage
有载调压　on-load voltage regulation
有载运行　on-load operation
淤泥　silt
余摆线　trochoid
余割　cosecant
余热锅炉　heat recovery boiler
余热回收　waste heat recovery
余弦　cosine
逾限　jabber
逾限控制　jabber control
与非电路　NOT AND circuit
预电离相互作用　ignition interaction
预防性检修　preventive maintenance
预防性维护　preventive maintenance

预计可靠度 predicted reliability
预烧测试 burn-in
预应力 prestress
域 field
阈电流 threshold current
阈电压 threshold voltage
阈值比 threshold ratio
元件 element
元件表 PL (parts list)
元素 element
元素成分 ultimate composition
元素分析 ultimate analysis
原电池 primary cell
原动机 prime mover
原位 home position
原型 prototype
原子 atom
原子电池 atomic battery
原子核 nucleus
原子裂变 atomic fission
原子碰撞 atomic collision
原子序数 atomic number
原子组成 atomic composition
圆角 fillet
圆球度 sphericity
圆筒形阀 cylindrical valve
圆锥 taper cone
圆锥铰刀 taper reamer
圆锥销 taper pin
远场衍射图样 far-field diffraction pattern
远程计算 telecounting
远程数据处理 TDP (teledata processing)
远动 telecontrol
远动跳闸 transferred tripping
远方跳闸 transferred tripping
约翰逊热噪声 Johnson noise
约瑟夫森效应 Josephson effect
跃迁 transition
跃迁能 transition energy
跃迁频率 jump frequency
跃迁时间 transition time
越带指令 tape skip
越线编码 out-of-line coding
云母电容器 mica capacitor
云母绝缘 mica insulation
匀速调节器 isodromic governor
允许中断 interrupt enable
运筹学 operational research
运输量 traffic
运输系数 transport coefficient
运输延迟单元 transport delay unit
运算 operation
运算放大器 operational amplifier
运行能力 serviceability
运行条件 service condition
运行转速 running speed
运转 operation
运转试验 in-service test
运转中断期 outage
熨平宽度 ironing width

匝间短路 interturn short circuit
匝间故障 interturn fault
匝间击穿试验 interturn breakdown test
匝间绝缘 interturn insulation
匝间试验 interturn test; turn-to-turn test
匝绝缘 turn insulation
匝链 linkage
匝数补偿 turn compensation
杂散损耗 stray loss
杂质 contaminant
载流量 ampacity
载体 bearer
载体催化元件传感器 sensor with supporter catalyst filled element
再沸器 reboiler
再燃烧 secondary combustion
再入控制 reentry control
再生能源 renewable energy
再生器 regenerator
在役检查 in-service inspection
暂时工作区 transient working area
暂态 transient state
暂态发电机 transient generator
暂态过电压 temporary overvoltage
暂态平衡 transient equilibrium
暂态蠕变 transient creep
暂态稳定(度) transient stability
暂态性强迫停运 transient-cause forced outage
早期事故 early failure
噪声等级 NR (noise rating)
噪声电压 noise voltage

噪声二极管 noise diode
噪声计 noise meter
噪声系数 NF (noise factor)
噪声限制器 noise clipper
噪声消减 muting
噪声源 noise source
噪声振幅 noise amplitude
择优取向 preferred orientation
增量编程 incremental programming
增量编译器 incremental compiler
增量测量方法 incremental measuring method
增量尺寸标注 incremental dimensioning
增量电阻 incremental resistance
增量故障概率 incremental probability of failure
增量控制 incremental control
增量模式 incremental mode
增量失效概率 incremental probability of failure
增量式进给 incremental feed
增量式指令 incremental command
增扭器 torque booster
增塑剂 plasticizer
增弹剂 elasticizer
增压变压器 booster transformer
增压流化床燃烧 pressured fluidized-bed combustion
增益 gain
增益边限 gain margin
增益波动 gain ripple
增益常数 gain constant
增益交越 gain crossover

增益交越频率 gain crossover frequency
增益均匀性 gain flatness
增益控制 gain control
增益控制器 gain controller
增益调节 gain adjustment
增益误差 gain error
增益系数 gain coefficient
增益斜率 gain slope
增益裕量 gain margin
轧辊 roller
闸 brake
闸刀开关 knife switch
闸刀式隔离端子 isolating blade terminal
闸门 gate; shuttle
闸室 lock chamber
斩波器 chopper
占线 LB (line busy)
栈 stack
栈存储器 stacked memory
张拉 stretching
张力 strain
张量磁导率 tensor permeability
张量磁化率 tensor susceptibility
章动 nutation
章动敏感器 nutation sensor
胀塑性 dilatancy
障碍物 obstacle
障碍物探测 obstacle detection
障碍衍射 obstacle diffraction
障碍增益 obstacle gain
着火 firing
着火电压 firing voltage
找正 adjustment
兆巴 mbar (megabar)
兆比特 megabit
兆电子伏(特) MeV (megaelectronvolt)
兆尔格 megaerg
兆乏 megavar
兆法拉 macrofarad; megafarad
兆伏安 megavolt-ampere
兆伏(特) Mv (megavolt)
兆高斯 megagauss
兆赫 MHz (megahertz)
兆焦(耳) megajoule
兆库仑 megacoulomb
兆米 megametre
兆欧 megohm
兆欧表 earthometer; megger; megohmmeter; tramegger
兆瓦(特) Mw (megawatt)
兆周 megacycle
兆字节 MB (megabyte)
照度计 light meter; luxmeter
照明电缆 lighting cable
照明电路 lighting circuit
照明电压 lighting voltage
照明负荷 lighting load
照明负载 lighting load
照明开关 lighting switch
照明馈路 lighting feeder
照明网络 lighting network
照明有效性因数 lighting effectiveness factor
折射率 refractive index
折射指数 refractive index
折向器 deflector
折焰角 furnace arch
针形阀 needle valve
针焰试验 needle flame test
帧间时间填充 interframe time fill
帧失调 out-of-frame
真空 vacuum
真空表 vacuum gauge
真空传感器 vacuum sensor; vacuum transducer
真空阀 vacuum valve
真空干燥箱 vacuum drying oven
真空管 electron tube
真空规 vacuum gauge
真空熔炼 vacuum melting
真空熔融 vacuum melting
真空系统 vacuum system
真空修正 vacuum correction
真空压力计 vacuum gauge
真空蒸发 vacuum evaporation
真空贮气筒 vacuum accumulator
枕盒 crib
振荡 oscillation

振荡闭锁 out-of-step blocking
振荡传输机 shuttle
振荡电弧 oscillating arc
振荡电路 oscillating circuit
振荡电压 oscillating voltage
振荡发生器 oscillation generator
振荡放电 oscillating discharge
振荡管 oscillation tube
振荡函数 oscillating function
振荡回路电容器 tank capacitor
振荡频率 oscillating frequency
振荡器 oscillator
振荡速调管 oscillating klystron
振荡线圈 oscillating coil
振荡效应 oscillation effect
振荡周期 oscillating period
振荡阻尼 oscillation damping
振动 oscillation; vibration
振动传感器 vibration sensor; vibration transducer
振动分析仪 vibration analyzer
振动计 vibrometer
振动监测仪 vibration monitor
振动监视器 vibration monitor
振动检流计 vibration galvanometer
振动控制仪 vibration controller
振动烈度 vibration severity
振动炉排 vibrating grate
振动膜 tympanum
振动黏度计 vibrating viscometer
振动器 vibrator
振动试验 vibration test
振动台 vibration table
振动误差 vibration error
振动样品磁强计 VSM (vibrating sample magnetometer)
振幅 (vibration) amplitude
振簧系仪表 vibrating reed instrument
振筒式气压表 vibration cylinder barometer
振筒式气压传感器 vibration cylinder pressure transducer
振弦式传感器 vibrating wire transducer
振弦式拉力计 vibrating wire drawing force meter
振弦式力传感器 vibrating wire force transducer
振弦式张力传感器 vibrating wire tension transducer
振弦式张力计 vibrating wire tensiometer
振弦式转矩测量仪 vibrating wire torque measuring instrument
振弦式转矩传感器 vibrating wire torque transducer
振子天线 element antenna
震底式炉 shaker hearth furnace
镇静钢 killed steel
镇流器 ballast
镇流器系数 ballast factor
蒸发冷却器 devaporizer
蒸汽 steam
蒸汽导管 steam lead
蒸汽发电厂 steam power plant
蒸汽管道 steam lead
蒸汽管路吹扫 steam line blowing
蒸汽锅炉 steam generator
蒸汽含水量 water content of steam
蒸汽喷射流 steam jet
蒸汽清洗 steam washing
蒸汽室 steam chest
蒸汽弯曲应力 steam bending stress
整步 synchronizing
整步电抗器 synchronizing reactor
整步迭代 total step iteration
整步绕组 synchronizing winding
整定 setting
整距绕组 full-pitch winding
整流 rectification
整流变压器 rectifier transformer
整流片 fairing
整流器 rectifier

整流罩 cowl; fairing
整体式开关装置 integral switching device
整形电路 waveshaping circuit
正长石 orthoclase
正常频率 normal frequency
正冲击响应谱 positive shock response spectrum
正电性 electropositivity
正反馈 positive feedback
正负三位作用 positive-negative three-step action
正负作用 positive-negative action
正光电导的 light-positive
正极柱 positive terminal
正交极 orthopole
正交极化 orthogonal polarization
正交矩阵 orthogonal matrix
正交扫描 orthogonal scanning
正交时间相位差 time quadrature
正交时间相移 time quadrature
正交性 orthogonality
正交轴 orthoaxis
正离子 cation
正切 tangent
正切标度 tangent scale
正切尺 tangent scale
正切电流计 tangent galvanometer
正切模量矩阵 tangent modulus matrix
正摄像仪 orthodiagraph
正视图 orthograph
正态化 normalization
正态性 normality
正投射法 orthography
正投影 orthography
正投影图 orthograph
正温度系数热敏电阻 positive temperature coefficient thermistor
正系统 positive system
正弦 sine
正弦变化 sinusoidal variation
正弦函数 sine function
正弦绕组 sine winding
正向击穿 forward breakdown
正压 positive pressure
正压外壳 pressurized enclosure
正应变 positive strain
正余弦旋转变压器 sine-cosine resolver
正轴 orthoaxis
支撑物 bolster
支撑柱 shore
支杆触头 prod
支杆法 prod method
支索 jackstay
支线 leg; tapped line
支座 bearer
执行器 actuator
直供电源 transformerless power supply
直角点阵 orthogonal lattice
直角通电法 vertical current flow method
直径 diameter
直流电 DC (direct current)
直流电动机 direct current motor
直流电子开关 electronic DC switch
直流感应 influence by DC
直线加速器 linear accelerator
直线塔 tangent tower
直线性 linearity
止漏环 sealing
止水 sealing
纸带穿孔(机) tape punch
纸带读入机 tape reader
指零测量 null measurement
指令 instruction
指令集 instruction set
指令码 instruction code
指令系统 instruction set
指令译码器 instruction decoder
指示 instruction
指示极值监视器 indicating limit monitor
指示器 indicated device

指示器盘 indicator board
指示熔断器 indicating fuse
指示值 indicated value
指数法 index method
指针 pointer
指针长度 pointer length
指针偏转 needle deviation
指针式检流计 pointer galvanometer
指针式仪表 pointer instrument
指针调整 pointer adjustment
指针指示器 needle indicator
制表 tabulation
制表机 tabulating machine; tabulator
制表卡片 tabulating card
制表系统 tabulating system
制导系统 guidance system
制动开关 tappet switch
制动器 arrester; brake
制冷 refrigeration
制热量 heating capacity
制约性 conditionality
质量 mass
质量电阻率 mass resistivity
质量管理体系 QMS (quality management system)
质量指标 QI (quality index)
质谱法 MS (mass spectrometry)
质谱学 mass spectroscopy
质谱仪 mass spectrograph; mass spectrometer
质心 barycentre
质子 proton
智能充电器 intelligent charger
滞后 lag
滞后补偿 lag compensation
滞后电流 lagging current
滞后负载 lagging load
滞后网络 lag network
滞后相 lagging phase
滞后组件 lag module
置换 displacement; transposition
中断 interruption
中断表 interrupt list
中断程序 interrupt routine
中断处理 interrupt processing
中断处理器 interrupt handler
中断处理系统 interrupt handling system
中断等级 interrupt level
中断堆栈 interrupt stack
中断服务程序 ISE (interrupt service routine)
中断功能 interrupt function
中断计时器 interrupt timer
中断控制器 interrupt controller
中断排列 interrupt arrangement
中断屏蔽 interrupt mask
中断器 interrupter
中断请求 interrupt request
中断确认 interrupt confirmation
中断时间 interrupting time
中断矢量 interrupt vector
中断矢量控制 interrupt vectoring
中断向量 interrupt vector
中断向量控制 interrupt vectoring
中断溢出 interrupt overflow
中断优先级 interrupt priority (level)
中断字溢出 interrupt word overflow
中轭 intermediate yoke
中规模集成电路 MSI (medium-scale integration)
中和 neutralization
中和剂 neutralizer
中继局 tandem office
中继选择 tandem selection
中间变压器 interstage transformer
中间测量 intermediate measurement
中间测试 interim test
中间层 interlayer; intermediate layer
中间齿轮箱 intermediate gear box
中间点 intermediate point

中间电极 intermediate electrode
中间电压端子 intermediate voltage terminal
中间端子 intermediate terminal
中间负荷 shoulder load
中间极 interpole
中间极绕组 interpole winding
中间开关 intermediate switch
中间冷却回路 intermediate cooling circuit
中间散热回路 intermediate cooling circuit
中间设备 intermediate equipment
中间线 mid-line
中间形态 intermediate form
中间支撑 intermediate support
中间轴 jack shaft
中间状态变量 internal state variable
中阶梯光栅 echelle grating
中介继电器 interposing relay
中介子 neutretto
中频 MF (medium frequency)
中频变压器 intermediate frequency transformer
中频放大器 intermediate frequency amplifier
中水回用 wastewater renovation
中微子 neutrino
中位数 median
中线 median
中线电流 neutral current
中心销 kingbolt
中性 neutrality
中性导体 neutral conductor
中性点 neutral point
中性点接地 neutral earthing; neutral grounding
中性点接地电抗器 neutral earthing reactor
中性点绝缘系统 isolated neural system
中性点有效接地系统 system with effectively earthed neutral
中性化 neutralization
中性接地制 grounded neutral system
中性母线 neutral bus
中性线 neutral line
中性线开关 neutral switch
中性相位 neutral phase
中压绕组 intermediate voltage winding
中央处理器 CPU (central processing unit)
中子 neutron
中子掺杂 neutron doping
中子发生器 neutron generator
中子密度 neutron density
终端过滤器 ultimate filter
终端盒 terminal box
终端-计算机多路转换器 TCM (terminal-to-computer multiplexer)
终端接口模件 TIM (terminal interface module)
终端控制系统 TCS (terminal control system)
终端区域分布处理 TADP (terminal area distribution processing)
终端塔 terminal tower
终端线 tag wire
终端站 headend
终检 final inspection
终接电路 final circuit
重心 barycentre
重油 maz(o)ut
周波变流器 cycloconverter
周线 contour
周缘流量 peripheral flow rate
轴 shaft
轴测法 axonometry
轴承 bearing
轴电流 shaft current
轴电压试验 shaft-voltage test
轴功 shaft work
轴环 collar
轴颈轴承 journal bearing
轴流式 axis-flow
轴流式水轮机 Kaplan turbine

轴流式通风机 axial fan
轴向聚焦 axial focusing
肘板 knee
骤冷阶段 quench cooling phase
骤冷前沿 quench front
骤冷箱 quench drum
主变压器 main transformer
主波瓣 major lobe
主从触发器 master-slave flip-flop
主从机械手 master-slave manipulator
主从式操作系统 master-slave operating system
主电弧 main arc
主电流 principal current
主电路 main circuit
主阀 king valve
主分接 principal tapping
主干 backbone
主回路 main circuit; major loop
主计算机 host computer
主绝缘 main insulation; major insulation
主控继电器 master relay
主控调速器 topping governor
主梁柱 king post
主令传动 master drive
主令开关 master switch
主螺栓 kingbolt
主频率 master frequency
主驱动电动机 master motor
主绕组 main winding
主线芯 master core
主芯片 master chip
主轴 spindle
主轴式天线 king-post antenna
助推器 booster
住宅与楼宇电子系统 HBES (home and building electronic system)
贮存寿命 shelf life
注入变压器 injection transformer
注入电流 injection current
注入电致发光 injection electroluminescence
注入电子束磁控管 injected-beam magnetron
注入法 injection method
注入剂量 injected volume
注入能量 injection energy
注入式激光二极管 injection laser diode
注入水平 injection level
注入同步范围 injection locking range
注射成型 injection moulding
注水 water injection
注塑 injection moulding
注塑件 injection part
注压件 injection part
驻波 clapotis
铸铁 cast iron
专用集成电路 ASIC (application specific integrated circuit)
专用说明书 SI (special instruction)
专用信息 SI (special information)
砖坯黏土 adobe
转变点 transformation point
转变能 transition energy
转变曲线 transition curve
转变温度 transition temperature
转换 transition
转换触点 transfer contact
转换点 inversion point
转换电压比 transfer voltage ratio
转换开关 COS (change-over switch)
转换器 converter; transducer
转接 transition
转接局 tandem office
转接开关 transfer switch
转接系统 intermediate system
转接选择 tandem selection
转接中继线 tandem-completing trunk
转录器 transcriber

转向架 bogie
转移常数 transfer constant
转移导纳 transfer admittance
转移电路 transfer circuit
转移指令 transfer order
转移阻抗 transfer impedance
转置积分方程 transposed integral equation
转置矩阵 transposed matrix
转罐式炉 rotary retort furnace
转矩波动 torque ripple
转矩补偿器 torque compensator
转矩传感器 torque pickup
转矩磁力计 torque magnetometer
转矩电流 torque current
转矩滑差特性 torque-slip characteristic
转矩计算 torque calculation
转矩限制 torque limitation
转轮 runner
转轮体 runner hub
转盘 turntable
转盘电极 rotating disc electrode
转速表 tachometer
转速表稳定系统 tachometer stabilized system
转速测定法 tachometry
转速继电器 tachometric relay
转子 armature; impeller; rotor
转子磁轭 rotor yoke
转子流量计〖带锥形管柱的〗 tapered tube rotameter
转子绕嵌机 rotor winding machine
转子支架 spider
装机容量 installed capacity
装配 assembly; installation
装配工 fitter
装配件 sub-assembly
装配器 assembler
装入程序 loader
装填 loading
装填机 packing machine
装卸机 loader and unloader
装卸桥 loader and unloader
装载机 loader
撞击器 striker
撞击强度 impact strength
撞击熔断器 striker fuse
追逐 hunting
锥塞阀 tapered plug valve
锥形负载电缆 taper-loaded cable
锥形滑阀 tapered slide valve
锥形级联 tapered cascade
锥形叶片 tapered blade
准分子离子 quasi-molecular ion
准峰值 quasi-peak
准峰值检波器 quasi-peak detector
准刚性转子 quasi-rigid rotor
准静不平衡 quasi-static unbalance
准平衡理论 quasi-equilibrium theory
准同步 quasi-synchronization
准线 directrix
准线性特性 quasilinear characteristics
准正弦量 quasi-sinusoidal quantity
准直管 collimator
准直仪 collimator
准周期振动 quasi-periodic vibration
着陆区投光灯 landing-area floodlight
子导线 sub-conductor
紫铜 copper
自保持 self-holding
自保持继电器 latching relay; lock-up relay
自焙电极 self-baking electrode
自查 self-inspection
自触发 self-triggering
自动测量系统 automatic measuring system
自动定时仪 time controller
自动断路器 automatic breaker
自动发电控制 AGC (automatic generation control)

自动幅值控制 AAC (automatic amplitude control)
自动跟踪控制 AFC (automatic following control)
自动关闭 self-closing
自动化孤岛 island of automation
自动还原电键 non-locking key
自动还原继电器 non-locking relay
自动给料机 self-feeder
自动给水控制 AFWC (automatic feed water control)
自动计算装置 ACE (automatic computing equipment)
自动切断器 electrotome
自动燃烧控制 ACC (automatic combustion control)
自动收报机 ticker
自动顺序 automatic sequence
自动同步发送器 transmitting selsyn
自动信息处理 automated information processing
自动遥控 ARC (automatic remote control)
自动制图 automated mapping
自复位 self-resetting
自攻螺钉 tapping screw
自攻螺纹 tapping screw thread
自换相 self-commutation
自恢复绝缘 self-restoring insulation
自加热 self-heating
自检 self-checking; self-inspection
自来水厂 waterworks
自冷式 self-cooled type
自励 self-excitation
自耦变压器 autotransformer; transtat; variac
自耦变压器起动器 autotransformer starter
自起动器 self-starter
自然阻尼 natural damping
自热 self-heating
自锁开关 non-homing switch
自同步机 selsyn
自位轴承 self-aligning bearing
自熄 self-extinguishing
自熄弧故障 self-extinguishing fault
自用变压器 house transformer
自用发电站 in-plant power station
自整角机 selsyn
自重-载重比 tare-load ratio
字段 field
字母码 alphabetic code
综合数据网 integrated data network
综合业务数字网 ISDN (integrated service digital network)
总安装功率 gross installed capacity
总波特率 gross baud rate
总地线 ground bus
总电负荷 total electric load
总电容 total capacitance
总发电量 gross generation
总阀 king valve
总幅值 total amplitude
总复位 general reset
总功率 gross output
总含盐量 TDS (total dissolved solids)
总结指令 tally order
总开关 incoming circuit-breaker
总流量 total flow
总落差 gross head
总热效率 gross thermal efficiency
总热值 total heating value
总溶解固体 TDS (total dissolved solids)
总受热面 total heating surface
总输出 gross output
总输入阻抗 total input impedance
总水分 total moisture
总水头 gross head; total head

总通量 total flux
总吸收率 total absorptivity
总压头 total head
总有机碳 TOC (total organic carbon)
总振幅 total amplitude
总装机容量 gross installed capacity
总阻抗 total impedance
纵横开关 cross bar
纵距 latitude
纵联保护 pilot protection
纵联复式汽轮发电机 tandem compound turbogenerator
纵联复式汽轮机 tandem compound turbine
纵列多丝埋弧焊 tandem sequence submerged arc welding
纵列配置 tandem arrangement
纵列绕组线圈 tandem winding coil
纵向负载 longitudinal load
纵向平直度 lengthwise flatness
纵向振荡 longitudinal oscillation
纵轴 y-axis
纵坐标 y-coordinate
阻碍 obstacle
阻挡物 obstacle
阻化剂 inhibitor
阻化油 inhibited oil
阻火器 flame arrester
阻聚剂 inhibitor
阻抗 impedance
阻抗比 impedance ratio
阻抗变换器 impedance converter
阻抗电压〖额定电流下的〗 impedance voltage at rated current
阻抗继电器 impedance relay
阻抗接地 impedance earthing
阻抗接地系统 impedance-earthed system
阻抗矩阵 impedance matrix
阻抗耦合 impedance coupling
阻抗平衡继电器 impedance balance relay
阻抗起动器 impedance starter
阻抗损耗 impedance loss
阻抗压降 impedance drop
阻尼 damping
阻尼器 damper
阻燃结构 flame-retarding construction
阻塞 congestion
组合装置 assembly
组件 assembly
组态 configuration
钻孔器 aiguille
钻模样板 jig
钻头 aiguille; bit
钻头柄 shank
最大持续功率 MCR (maximum continuous rating)
最大冲击电压 impulse crest voltage
最大出力 ultimate output
最大负荷 maximum load; ultimate load
最大功率 ultimate capacity
最大连续定额 MCR (maximum continuous rating)
最大容量 maximum capacity; ultimate capacity
最大输出 ultimate output
最大增益 ultimate gain
最大振幅滤波器 MAF (maximum amplitude filter)
最大装机容量 ultimate installed capacity
最低测得频率 LOF (lowest observed frequency)
最高电压 maximum voltage
最高负荷 maximum load
最高灵敏度 ultimate sensibility; ultimate sensitivity
最高设备电压 highest voltage for equipment
最后抽样单位 ultimate sampling unit
最后贮存桶 ultimate storage drum

最佳参数　optimal parameter
最佳耦合　optimum coupling
最佳判据　optimal criterion
最佳配置　optimum allocation
最佳输出　optimum output
最佳效率　optimal efficiency
最佳阻尼　optimal damping
最小点燃电流　MIC（minimum igniting current）
最小分断电流　minimum breaking current
最小值　minimum value
最优分配　optimum allocation
最优化　optimization
最优控制　optimal control
最优控制器　optimizer
最终被控变量　ultimately controlled variable
最终产量　ultimate yield
最终处理　ultimate disposal
最终处置　ultimate disposal
最终检查　final inspection
最终检验　final inspection
最终燃料消耗　ultimate fuel burn-up
最终热阱　ultimate heat sink
最终热阱水坑　ultimate heat sink basin
最终受控变量　ultimately controlled variable
最终温度　ultimate temperature
最终压力　ultimate pressure
最终贮存　ultimate storage
左扭转　left-handed twist
左手定则　LHR（left-hand rule）
左向旋转　left-handed rotation
左旋　left-handed rotation
左转螺旋　left-handed screw
作业处理　job processing
作业调度　job scheduling
作业队列　job queue
作业分解结构　WBS（work breakdown structure）
作业数据　job data
坐标　coordinate
坐标网点　grid point
做功冲程　power stroke

非汉字开头的词条

APT 语言　APT (Automatically Programmed Tool)
D 控制器　D-controller; derivative controller
D 作用　D-action; derivative action
FAMOS 存储器　FAMOS memory
H 矢量　H-vector
H 型电缆　H-type cable
ISO 基准公差系列　ISO fundamental tolerance series
IT 系统　IT system
K 波段　K-band
K 电子　K-electron
K 俘获　K-capture
K 壳　K-shell
K 频带　K-band
K 系　K-series
K 型架构　K-frame structure
K 型显示　K-display
K 型载波系统　K-carrier system
K 转换　K-conversion
L 分级器　L grader
L 形天线　L aerial; L antenna
L 形网络　L-network
L 型阴极　L-cathode
O 形垫圈　O-ring gasket
O 形环　O-ring
O 形环密封　O-ring seal
O 型管　O-type tube
O 型器件　O-type device
O 型显示器　O-scope
pH 指示器　pH-indicator
p 沟(道)场效(应)晶体管　p-channel FET
P 型半导体　hole semiconductor; p-type semiconductor
P 型掺杂剂　p-type dopant
P 型掺杂漏极　p doped drain
P 型掺杂源极　p doped source
P 型接触　p contact
P 型扩散区　p diffused region
P 作用　P-action
SI 单位　SI unit
S 形曲颈管　S-shaped trap
T 形电路　T-circuit
T 型接头　T joint
T 型衰减器　T-attenuator
T 型天线　T-type antenna
U 形钩　clevis
U 型磁铁芯　U core
U 型管　U-tube
U 型管压力计　U-tube manometer
U 型火焰炉　U-flame furnace
U 型夹　U-clamp
U 型螺栓　U-bolt
U 型弯曲试样　U-bend specimen
W 型电桥　W-bridge
W 型光纤　W-type fibre
X－Y 记录器　X-Y recorder
X 蜡　X-wax
X 频带　X band
X 射线管电流　X-ray tube current
X 射线机　X-ray machine
X 射线激光器　XRL (X-ray laser)
X 射线全身辐照　whole-body X-irradiation
X 射线衍射　XRD (X-ray diffraction)
X 射线荧光检查仪　XRF (X-ray fluoroscope)

X 形网络　X-network
X 型电桥　X-bridge
x 轴　x-axis
x 轴方向　x-direction
x 轴信号放大器　x-axis amplifier
x 坐标　x-coordinate
Y 环行器　Y circulator
Y 级绝缘材料　Y-class insulation
Y 接法　Y-connection
Y 接法电压　wye voltage
Y 接法相绕组　Y-connection phase winding
Y 接头　Y-junction
Y 切割　Y-cut
Y 形拉线　Y-stay
Y 形连接电阻箱　Y box
Y 形匹配　Y-matching
Y 形支架　wye
Y 型接头　Y joint
Y 型铁芯　Y-type core
Y 型雾化器　Y-jet atomizer
Y 型油喷燃器　Y-jet type oil burner
y 轴　y-axis
y 轴方向　y-direction
y 轴线信号放大器　y-axis amplifier
Z 变换　Z-transform
Z 形联结　zigzag connection
Z 型钢　Z-section steel
Z 型角铁　Z-angle
z 域　z-domain

2,3-二羟基丁二酸　tartaric acid

α 射线　alpha radiation
Δ-Δ(接法)　twin-delta

参考文献

北勾等.燃料电池模拟、控制和应用.刘通译.北京:机械工业出版社,2011.

伯茨那编.英德汉电工技术与自动化词典.林绳宗等译.北京:机械工业出版社,2002.

程志光等.电气工程电磁热场模拟与应用.北京:科学出版社,2009.

戴庆忠主编.英汉电技术词汇.第2版.北京:中国电力出版社,2007.

德马雷斯特.工程电磁学.影印本.北京:科学出版社,2003.

《电气工程师(供配电)实务手册》编写组编.电气工程师(供配电)实务手册.北京:机械工业出版社,2006.

《电气工程师手册》第2版编辑委员会编.电气工程师手册.第2版.北京:机械工业出版社,2000.

丁坚勇.电与电能.北京:中国电力出版社,2008.

冯垛生主编.太阳能发电原理与应用.北京:人民邮电出版社,2007.

《工厂常用电气设备手册》编写组编.工厂常用电气设备手册(上册).第2版.北京:中国电力出版社,1997.

韩至成等.太阳能级硅提纯技术与装备.北京:冶金工业出版社,2011.

何伯述主编.热能与动力机械基础.北京:清华大学出版社,北京交通大学出版社,2010.

胡红光编.电力设备红外诊断技术与应用.北京:中国电力出版社,2012.

胡念苏主编.发电动力系统概论.北京:中国水利水电出版社,2011.

胡钋,华小梅.电力工程英语综合教程.北京:中国电力出版社,2011.

靳瑞敏.太阳能电池原理与应用.北京:北京大学出版社,2011.

孔力主编.英汉电工电力词汇.北京:科学出版社,2007.

李鹏飞主编.电力电子技术与应用.北京:清华大学出版社,2012.

连国钧.动力控制工程.西安:西安交通大学出版社,2001.

廖晓昕.动力系统的稳定性理论和应用.北京:国防工业出版社,2000.

刘然等.电力专业英语.第2版.北京:中国电力出版社,2004.

罗宾逊.动力系统导论(英文版).北京:机械工业出版社,2005.

罗利文等.电气与电子测量技术.北京:电子工业出版社,2011.

单文培等主编.电气设备试验及故障处理实例.第2版.北京:中国水利水电出版社,2012.

上海电力学院外语系组编.新编电力英语教程(上卷).北京:电子工业出版社,2004.

上海电力学院外语系组编.新编电力英语教程(中卷).北京:电子工业出版社,2004.

上海电力学院外语系组编.新编电力英语教程(下卷).北京:电子工业出版社,2004.

上海市电力公司、超高压输变电公司组编.电气试验.北京:中国电力出版社,2011.

舒飞等.AutoCAD 2005电气设计.北京:机械工业出版社,2005.

瓦格曼,艾施里希.太阳能光伏技术.叶开恒译.西安:西安交通大学出版社,2011.

汪光裕.光伏发电与并网技术.北京:中国电力出版社,2010.

王长贵,王斯成主编.太阳能光伏发电实用技术.第2版.北京:化学工业出版社,2009.

王海云等.风力发电基础.重庆:重庆大学出版社,2010.

王华忠等.电气与可编程控制器原理及应用.北京:化学工业出版社,2011.

王建华主编.电气工程师手册.第3版.北京:机械工业出版社,2006.

王其红等主编.电工手册.郑州:河南科学技术出版社,2010.

王为民等.核能发电与核电厂水电热联产技术.北京:化学工业出版社,2008.

王志新.现代风力发电技术及工程应用.北京:电子工业出版社,2010.

吴佳梁等.风力机可靠性工程.北京:化学工业出版社,2010.

西安交通大学《英汉电工电子技术词典》编写组编.英汉电工电子技术词典.第2版.北京:机械工业出版社,2000.

西安热工研究院.发电企业节能降耗技术.北京:中国电力出版社,2010.

谢亚青,郝春玲主编.液压与气动技术.上海:复旦大学出版社,2011.

徐铭陶,肖明葵.工程动力学振动与控制.北京:机械工业出版社,2004.

许珉等编.发电厂电气主系统.第2版.北京:机械工业出版社,2011.

姚兴佳,宋俊.风力发电机组原理与应用.第2版.北京:机械工业出版社,2011.

叶予光,王辉主编.电力电子技术.北京:中国电力出版社,2011.

易大贤主编.发电厂动力设备.第2版.北京:中国电力出版社,2008.

《英汉电工电子技术词典》编写组编.英汉电工电子技术词典.北京:机械工业出版社,2000.

《英汉汉英电力词典》编委会编.英汉汉英电力词典.北京:中国电力出版社,2008.

约斯特.动力系统.影印版.北京:科学出版社,2006.

赵阳主编.电气与电子工程专业英语.第2版.北京:机械工业出版社,2009.

张友汉主编.电气技师手册.福州:福建科学技术出版社,2004.

郑体宽主编.热力发电厂.第2版.北京:中国电力出版社,2008.

中国动力工程学会主编.火电站系统与辅机.火力发电设备技术手册:第4卷.北京:机械工业出版社,2002.

中国科学技术协会,中国电机工程学会.2009—2010动力与电气工程学科发展报告.北京:中国科学技术出版社,2010.

外教社英汉·汉英百科词汇手册系列

1. 保险
2. 材料学
3. 财政学
4. 测绘学
5. 出版印刷
6. 地理学
7. 地球科学
8. 地质学
9. 电子、通信与自动控制技术
10. 电子商务
11. 动力与电气工程
12. 法学
13. 房地产
14. 纺织
15. 工程与技术
16. 管理学
17. 广播影视
18. 广告学
19. 国际贸易
20. 海洋与水文
21. 航空航天
22. 核科学技术
23. 化工
24. 化学
25. 环境科学
26. 机械工程
27. 计算机
28. 交通运输工程
29. 教育学
30. 经济学
31. 军事
32. 考古学
33. 会计与审计
34. 矿山工程技术
35. 力学
36. 历史学
37. 林业
38. 旅游
39. 美术、书法与摄影
40. 民族学
41. 能源科学技术
42. 农业
43. 烹饪
44. 人口学
45. 人类学
46. 人力资源管理
47. 商务
48. 社会学
49. 生理学
50. 生物学
51. 食品科学
52. 市场营销
53. 市政工程
54. 数学
55. 水产业
56. 水利工程
57. 体育
58. 天文学
59. 统计学
60. 图书馆、情报与文献学
61. 土木建筑
62. 网络与多媒体
63. 微生物与病毒学
64. 文学
65. 物理学
66. 物流与仓储
67. 戏剧、戏曲与舞蹈
68. 心理学
69. 新闻与传播
70. 信息与系统科学
71. 畜牧与兽医
72. 药学
73. 医学
74. 音乐
75. 语言学
76. 园艺学
77. 哲学
78. 政治学
79. 知识产权
80. 植物学
81. 中医中药
82. 宗教